For my son Nishil

TABLE OF CONTENTS

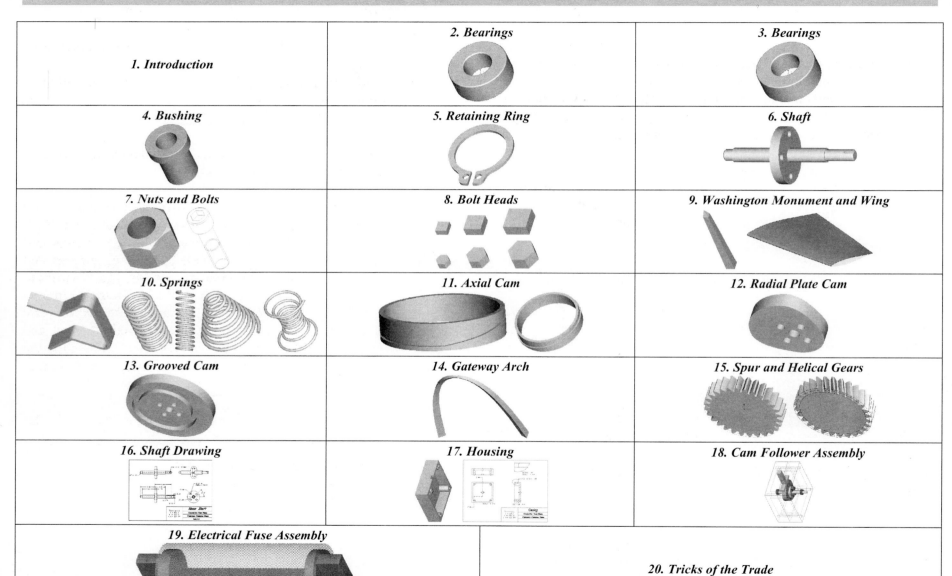

ML
edM

MECHA... ...DELING

McGraw Hill

Boston   Burr Ridge, IL   Dubuque, IA   Madison, WI   New York   San Francisco   St. Louis
Bangkok   Bogotá   Caracas   Kuala Lumpur   Lisbon   London   Madrid   Mexico City
Milan   Montreal   New Delhi   Santiago   Seoul   Singapore   Sydney   Taipei   Toronto

McGraw-Hill Higher Education

*A Division of The **McGraw-Hill** Companies*

MECHANICAL DESIGN MODELING USING PROENGINEER

International   1 2 3 4 5 6 7 8 9 0 QPD/QPD 0 9 8 7 6 5 4 3 2 1
Domestic   1 2 3 4 5 6 7 8 9 0 QPD/QPD 0 9 8 7 6 5 4 3 2 1

ISBN 0–07–244314–6
ISBN 0–07–112185–4 (ISE)

General manager: *Thomas E. Casson*
Publisher: *Elizabeth A. Jones*
Sponsoring editor: *Jonathan Plant*
Marketing manager: *Ann Caven*
Project manager: *Jane E. Matthews*
Production supervisor: *Sherry L. Kane*
Coordinator of freelance design: *David W. Hash*
Cover designer: *Emily Feyen*
Senior supplement producer: *David A. Welsh*
Media technology senior producer: *Phillip Meek*
Compositor: *Lachina Publishing Services*
Typeface: *11/13 Times New Roman*
Printer: *Quebecor World Dubuque, IA*

"The following are registered trademarks of Parametric Technology Corporation: Pro/ENGINEER, Pro/MECHANICA, Pro/INTRALINK, and all other applications in the Pro/ENGINEER family of products."

Library of Congress Cataloging-in-Publication Data

Condoor, Sridhar S., 1967–
   Mechanical design modeling using ProEngineer / Sridhar S. Condoor. — 1st ed.
     p.   cm.
   ISBN 0–07–244314–6 — ISBN 0–07–112185–4 (ISE)
    1. Engineering design—Data processing. 2. Pro/ENGINEER. 3. Mechanical drawing.
   4. Computer-aided design. I. Title.

   TA174 .C5875 2002
   620′.0042′02855369—dc21   2001044145
                               CIP

www.mhhe.com

Preface

Over the past several years, I taught project-based design courses to mechanical and aerospace engineering students. During the course of the projects, students found great difficulties in creating "real" parts and assembling them. Most books fall short in providing the reader with a consistent and systematic methodology for approaching even simple solid modeling tasks.

Graphics books do not deal with the solid modeling software in detail. On the other hand, solid modeling books do not handle both graphics and design topics well. This book is aimed at addressing this need and is based on the lecture notes developed to teach ProEngineer, graphics and also, aspects of design.

The focus of the text is on teaching actual design using ProEngineer rather than teaching a set of commands. The book illustrates the part, drawing and assembly creation with several industrial illustrative parts. These parts fit together in the final chapters to form one large assembly.

Chapters are organized such that each chapter builds on previous chapters and introduces additional commands. It is a hands-on book where students are expected to work with ProEngineer. Several illustrative pictures are used to illustrate each step. The book eliminates the frequent sight of students staring at the book and desperately trying to follow the instructions.

Mechanical Design Modeling Using ProEngineer is a book for graphics and design courses from freshmen- to senior-level students in mechanical and aerospace engineering. Practicing engineers will also find this book valuable. This book is aimed at the new generation of students who are:

- Highly computer literate (sometimes more literate than faculty).
- Not motivated in reading large volumes of information to do simple things.
- Hands-on when it comes to computers.

I hope this book will open the doors of the wonderful world of designing using solid modeling techniques.

I would like to thank the individuals who contributed to my design education, particularly Prof. Christian Burger and Prof. David Jansson. I would like to thank faculty, staff, students and friends at Saint Louis University for their continuous support and cooperation. Special thanks to two individuals Mr. Mark Kurfman and Mr. Larry Boyer, who contributed significantly in the development and testing of this book. I would like to express my appreciation to Jonathan Plant, Jane Matthews, and editorial and production staff at McGraw-Hill. Also, I wish to acknowledge the industrial support from Parametric Technology Corporation, Cooper Bussmann and Ferguson Industries.

Thanks to the following reviewers for their excellent feedback:
Kent Lawrence, University of Texas at Arlington
Mike Philpott, University of Illinois
John Renaud, University of Notre Dame
C. Steve Suh, Texas A&M University

About the Author

Sridhar S. Condoor received his Ph.D. from the Department of Mechanical Engineering at Texas A&M University. He received M.S. from the Indian Institute of Technology, Bombay and B.S. from Jawaharlal Nehru Technological University, Hyderabad, India. His research interests include many areas of Design Theory and Methodology, Computer Aided Design, Cognitive Science and Mechatronics.

Condoor is the co-author of *Innovative Conceptual Design: Applications: Theory and Application of Parameter Analysis* (Cambridge University Press). He conducted several short-courses for both faculty and practicing engineers on design techniques. He coauthored several papers on various facets of design including design management, design theory, design principles, cognitive aspects of design and design education.

Condoor currently teaches at Saint Louis University – Parks College of Engineering & Aviation. He is initiated the multidisciplinary design program with a strong emphasis on industry participation at Saint Louis University. He works with a diverse range of industries in creating new products, modeling existing products to understand the underlying physics and developing automated manufacturing systems.

LESSON 1
INTRODUCTION

What is parametric, feature-based design?

Parametric Technologies Corporation (PTC) revolutionized the CAD industry in the late 1980s with the introduction of ProEngineer, the first parametric, feature-based solid modeling CAD system with a strong emphasis on design intent. As opposed to 2-D drafting and 3-D Boolean modeling techniques, feature-based design is very intuitive. A part is created using features such as protrusions, cuts, holes, chamfers and rounds. The feature names can correspond to the actual physical features of the part. For instance, bolt head, socket, shank, threads and chamfers are the five key features of a hexagonal socket head bolt (Refer Fig. 1.1).

Features can be classified into *sketched* and *referenced* features. A sketched feature is created by sketching a 2-D section (sometimes several sections), and then moving the section along a predetermined path. For instance, the hexagonal socket is created by drawing a hexagon, extruding it along the axis up to a predetermined depth, and then removing the material contained within the extruded volume. Referenced features are pick-and-place features with predetermined shapes. Holes, rounds and chamfers are the three most commonly used referenced features. They involve selecting the references such as edges to be chamfered.

The term *parametric design* refers to the modeling technique wherein the design features, parts and assemblies are based on parameters whose values determine the physical shape of the design. Modifying the values of a particular parameter changes the corresponding feature and all the features that are referenced to this feature. For instance, if we increase the length of the shank, then the threads move along with it such that the threads start at the bottom of the shank. The key advantage of the parametric, feature-based design is the designer's ability to capture the *design intent* in the model.

ProE provides a practical way to develop flexible models that are easy to customize to suit the specific needs of the customers. Structural and thermal analysis, and motion simulation, can be carried out using ProMechanica in an integrated environment. The analysis often leads to greater understanding of the design and helps in identifying product improvements.

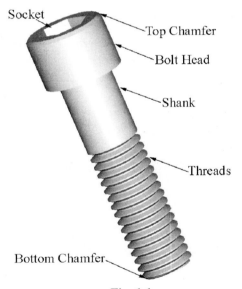

Fig. 1.1.

What is a parent-child relationship?

Feature-based design continuously creates child features, which depend on parent features for their existence. If the parent feature is deleted, then the user must delete, redefine, rollback or reroute its child features. Therefore, it becomes very important to make sure that the parent features exist throughout the modeling process.

Let us look at a simple example involving the positioning of a hole. The hole can be positioned (or dimensioned) with respect to two edges (Refer Fig. 1.2). The hole is a child feature and its existence depends on the parent features (edges 1 and 2). ProE prompts an error message if the edges are chamfered later in the design process. On the other hand, if the hole was referenced to surfaces 1 and 2 instead of the edges, then chamfering the edges would not prompt an error message.

TRICK: Referencing a feature with respect to surfaces will eliminate most parent-child problems.

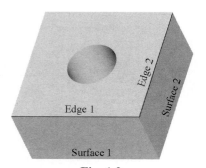

Fig. 1.2.

What is associative?

In ProE, the user can work in several modes. Some of the key modes are:

Sketcher – for creating sections and sketches
Part – for modeling parts
Drawing – for creating engineering drawings
Assembly – for assembling parts

A designer can create a sketch, use the sketch to model a part, create engineering drawing of the part, and also use the part in an assembly. The designer can modify the dimension of the part in any one of these modes. ProE handles this situation very well as all modes in ProE are fully associative (Refer Fig. 1.3). In other words, changes in one mode will automatically propagate to the other modes. Thus, it maintains consistency of the model.

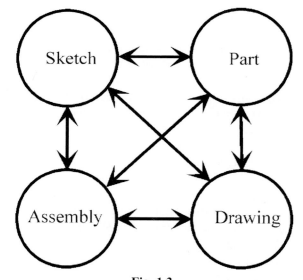

Fig. 1.3.

What are the main components of ProE display?

As shown in Fig. 1.4, the ProEngineer display consists of:

- **Main graphics window:** Displays the model.
- **Model tree window:** Lists the features and the order of their creation.
- **Menu manager:** Manages menus.
- **Pull down menus:** Provide several commands.
- **Top and Right toolbars:** Show shortcut icons to commonly used commands. These toolbars can be customized.
- **Command description window:** Provides concise description of the command on which the cursor is placed.
- **Message window:** Provides feedback to the user, and also directs the user while creating a feature. The previous messages can be viewed by using the scroll bar. The window can be resized by dragging the lower edge of the window.

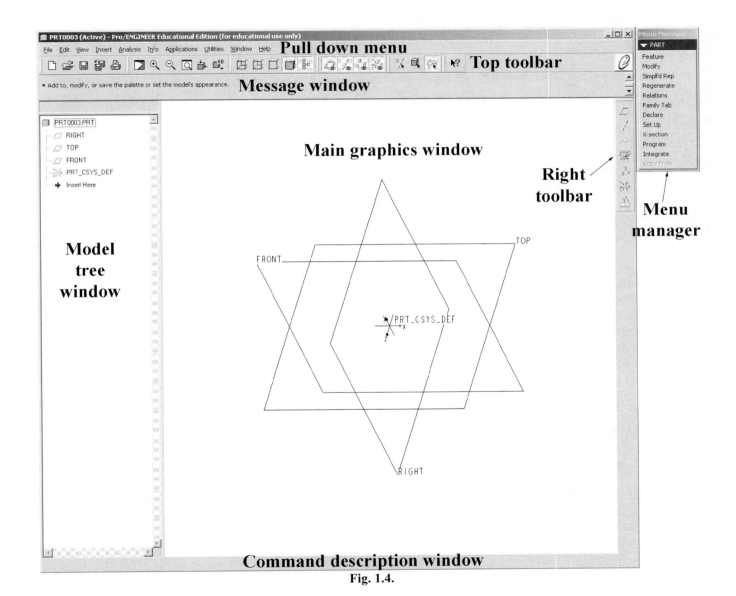

Fig. 1.4.

What is a good modeling process?

A good modeling process provides a general direction on how to approach the problems. The book recommends a three-step process as shown in Fig. 1.5. The steps are:

- **Step 1:** Select the command after reading its description.
- **Step 2:** Input the required information in the main graphics window.
- **Step 3:** Read the message window. ProEngineer confirms the command execution and also directs the user for further inputs.

While modeling, the focus of the eyes should continuously cycle through the menu manager, the command description window, the graphics window, and then the message window. The cursor, on the other hand, should move between the menu manager and the graphics window.

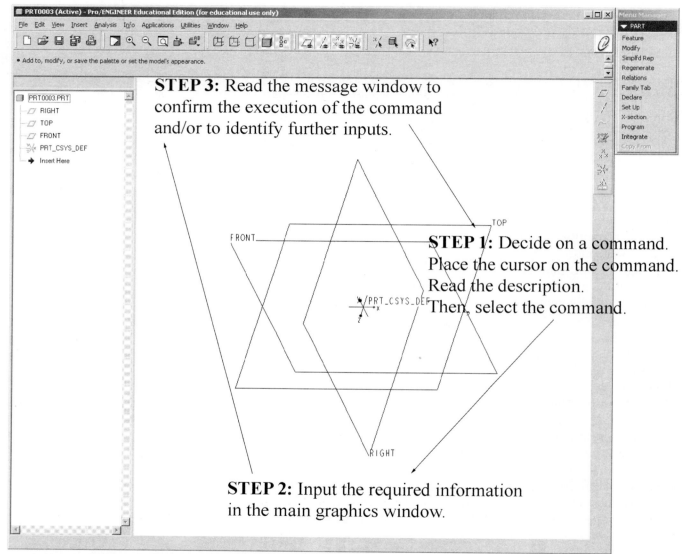

STEP 3: Read the message window to confirm the execution of the command and/or to identify further inputs.

STEP 1: Decide on a command. Place the cursor on the command. Read the description. Then, select the command.

STEP 2: Input the required information in the main graphics window.

Fig. 1.5.

Organization of the book

This book is heavily dependent on examples to elucidate the fundamental concepts in solid modeling in general and ProEngineer in particular. The examples are carefully detailed in a way that clarifies the thought process undertaken and the actual commands used. The author has seen firsthand the value of the "coaching method" that takes the reader from modeling simple machine components such as bearings, to complex components such as helical gears, and then to the bottom-up and top-down assembly approaches. The examples are chosen to exemplify engineering design characteristics such as equations, data points and graphs.

The reader is encouraged to work with ProEngineer and follow the modeling process throughout the book. The combination of "theoretical" discussion of the solid modeling principles and the "hands-on" ProE exercises has proved to be an effective method for this material to be learned.

Chapters 2 – 15 introduce the basic concepts in part modeling. Chapters 16 and 17 show how to create professional drawings. Chapters 18 and 19 show the bottom-up and top-down design approaches respectively. The book concludes with Chapter 20, which provides several tricks of the trade. The chapters are organized so that each chapter adds new knowledge about solid modeling and reinforces the previous chapters. Thus, the readers can sharpen their skills while acquiring new ones. This continuous reinforcement of concepts is one of the key features of this book.

Each chapter is organized into three sections. The first section provides a background about the part/assembly and information about the functionality and the design intent. The second section shows the sequence of steps involved in modeling the part/assembly. Even though the book shows one modeling approach, the reader is encouraged to explore alternative approaches for creating the same part and identify their advantages and disadvantages. Then, the third section provides detailed procedure for creating the part/assembly in ProE.

The procedure for creating a part/assembly is organized in terms of goals (describe major objectives), steps (describe steps involved in satisfying the goal) and commands (actual ProE commands). Several illustrative figures are included on each page to assist the reader in modeling. The commands are coded to provide fast access to the commands. Fig. 1.6 describes the command codes.

Command Codes

UPPERCASE LETTERS – Main menu.

Lowercase Letters – Menu Manager.

Underlined Letters – To be typed by the user.

Italics – Select or click with mouse.

BOLD LETTERS – Information about ProE.

Default Option – Default option in the menu.

BUTTON – Menu button.

KEYBOARD – Keyboard entry (typically either delete or enter key).

★ – Tricky step or read the instructions carefully.

ICON – Click on the icon.

ICON>ICONS – Click on > to see the icons.

Fig. 1.6.

REMEMBER TO SAVE THE PARTS.

MOST OF THE PARTS FIT TOGETHER
TO FORM THE ASSEMBLY IN LESSON 18.

The required lessons are:
- Lesson 2 – Bearing
- Lesson 6 – Shaft
- Lesson 7 – Nuts and Bolts
- Lesson 12 – Plate cam
- Lesson 16 – Shaft Drawing
- Lesson 17 – Housing

Learning Objectives:

- Understand the concept of **datum planes**.

- Explore the use of **mouse** for **zoom**, **spin**, and **pan** functions.

- Create **Protrusion – Extrude** and **Round** features.

Design Information:

Bearings allow relative motion between two components while minimizing frictional losses. For instance, the main bearings in an automobile allow the wheels to rotate relative to the body. Rolling element or anti-friction bearings are one of the common types of bearings. They consist of an outer race and an inner race separated by rolling elements (balls or cylinders). The rolling elements reduce friction by providing rolling contact. As bearings are purchased items, only the outer profile is modeled. Typically, rolling element bearings are mounted using interference fit. Therefore, the inner and outer diameters of the bearing are critical dimensions. For proper assembly, the edges of the bearing are rounded. The radius of the round is another critical dimension.

Sequence of Steps

Goal I: Experiment with the mouse

"Ctrl" Key +		
Left Mouse	**Middle Mouse**	**Right Mouse**
Zoom	Spin	Pan

Goal II: Understand datum planes

1. FRONT, TOP and RIGHT are the three default datum planes.

2. PRT_CSYS_DEF is the default coordinate system.

3. Spin center (Red, Green and Blue lines) helps in rotating the part.

Goal III: Create the base cylinder

1. Define the direction of extrusion.

2. Sketch two circles on the TOP datum plane.

3. Define the depth of extrusion.

Goal IV: Round the edges of the bearing

1. Select the edges to be rounded.

2. Specify the radius.

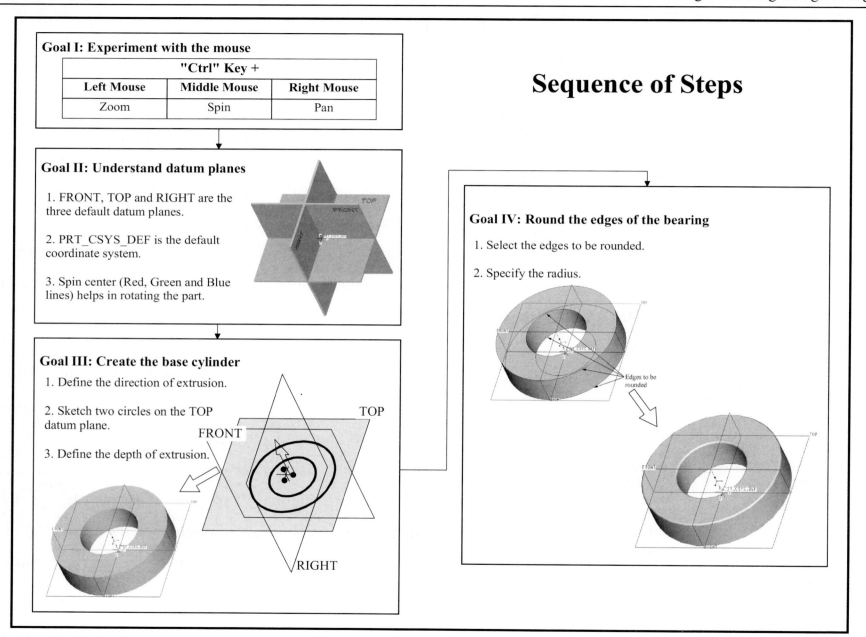

Goal	Step	Commands
Open a new file for the bearing part	1. Set up the working directory.	FILE → SET WORKING DIRECTORY → *Select the working directory* → OK **ProE always saves the file in the selected working directory.**
	2. Open a new file.	FILE → NEW → *Part* → *Solid* → bearing → OK **We will create the bearing as a solid part. ProE displays the three default datum planes in the graphics window.** **Refer Fig. 2.1.**
Experiment with the mouse	3. Use the mouse to zoom, spin and pan the model.	**In ProE, the mouse is a very powerful tool. We can zoom, spin and pan the model by holding the *CTRL* (Control key) and one of the mouse buttons down at the same time, and moving the mouse. The table in Fig. 2.2 illustrates the mouse functions. To get back to the default view, use the following command:** VIEW → DEFAULT ORIENTATION. **The default view is typically set as trimetric. However, it can be changed to isometric or user-defined by using** VIEW → REORIENT → (Type) *Preferences* **command.**

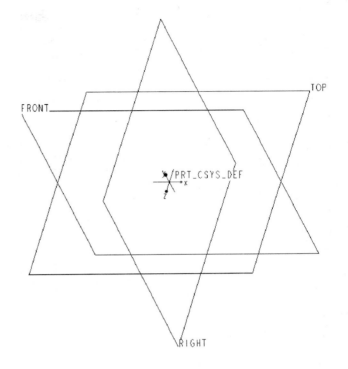

Fig. 2.1.

CTRL +		
Left Mouse	**Middle Mouse***	**Right Mouse**
Zoom	Spin	Pan

* If the middle mouse button is not working, use Shift key + Left mouse button for the middle mouse button.

Fig. 2.2.

Goal	Step	Commands
Understand the datum planes	4. Understand the datum planes.	★ **ProE creates three default datum planes named as FRONT, TOP and RIGHT. Each datum plane has two sides marked by two different colors: YELLOW and RED. In the default view shown in Fig. 2.1, only the yellow sides are visible. We can notice the two colors when we rotate the datum planes. We can visualize the planes better by looking at Fig. 2.3 where the planes are shaded.** **In Figs. 2.1 and 2.3, we can also see the default coordinate system "PRT-CSYS-DEF" at the center. The spin center is shown in Red, Green and Blue (RGB) color lines representing X, Y and Z axes respectively. The spin ceter helps in rotating the part.**
Create the base cylinder	5. Start "Protrusion – Extrude" feature.	Feature → Create → Solid → Protrusion → Extrude → Solid → Done
	6. Define the direction of extrusion.	One Side → Done **"One side" option extrudes the section by the specified length in one direction. "Both sides" extrudes the section in opposite directions.**

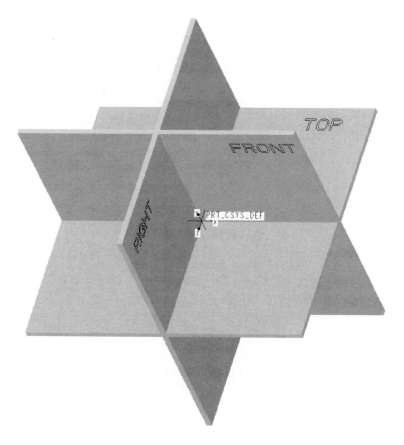

Fig. 2.3.

Goal	Step	Commands
	7. Select the sketching plane.	**We are going to sketch the section on the TOP datum plane.** Setup New → Plane → Pick *Select the TOP datum plane by clicking on the word "TOP"* → **Refer Fig. 2.4.** **The red arrow points to the direction of feature creation.** Okay
Create the base cylinder (Continued)	8. Orient the sketching plane.	**Specify a horizontal or vertical reference to orient (rotate) the sketching plane. As we are dealing with a symmetric object in this design task, the choice of the datum plane is not important.** Right → Plane → Pick → *Select the RIGHT datum plane*

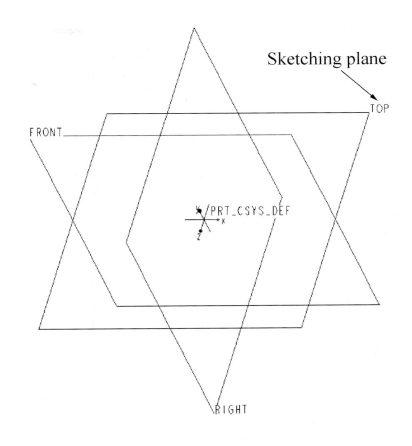

Sketching plane

TOP

FRONT

/PRT_CSYS_DEF

RIGHT

Fig. 2.4.

Orienting the sketcher:
In general, determine the intended direction of the outward normal from any datum plane (other than the sketching plane) in the sketch view. The outward normal originates from the yellow side. Select that direction from the SKET VIEW menu. Then, select the corresponding datum plane.

Goal	Step	Commands
Create the base cylinder (Continued)	9. Understand the orientation of the sketcher.	**The screen changes to the sketcher mode. It is advisable to rotate the model to understand where we are sketching.** ***CNTRL*** *+ Middle Mouse →* **To get back to the previous view, use the following command:** VIEW → SKETCH VIEW Or  → **The "References" window has two references: F1(RIGHT) and F3(FRONT).** **Refer Fig. 2.5.** **If this window is not visible, select** SKETCH → REFERENCES **All dimensions are placed with respect to the references. If necessary, additional references can be added to this list. It is advisable to select all the references before sketching.**

Fig. 2.5.

Goal	Step	Commands
Create the base cylinder (Continued)	10. Draw an outer circle.	○ → *Select the center of the circle as the intersection of the FRONT and RIGHT datum planes* → **Refer Fig. 2.6.** **The cursor snaps onto the intersection.** *Select a point to define the outer edge of the circle* **ProE automatically puts the dimension for the circle.** **Refer Fig. 2.6.**
	11. Create an inner circle.	○ → *Select the center of the circle as the intersection of the FRONT and RIGHT datum planes → Select a point to define the inner circle* **Refer Fig. 2.7.**
	12. Modify the dimensions.	↖ → *Double click the inner circle dimension →* 1→ ***ENTER*** *→ Double click the outer circle dimension →* 2 → ***ENTER*** **ProE automatically regenerates the section.**

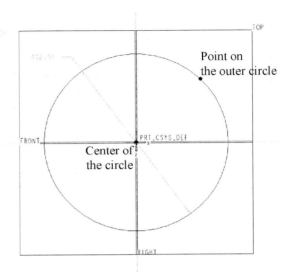

Fig. 2.6.

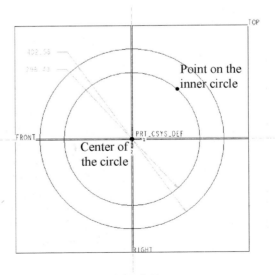

Fig. 2.7.

Goal	Step	Commands
Create the base cylinder (Continued)	13. Exit sketcher.	✔
	14. Define the depth.	Blind → Done → 0.5 → ✔ **In the "Blind" option, the user specifies the depth. The depth dimension can be modified later.**
	15. Accept the feature creation after previewing.	**Fig. 2.8 shows information about various element properties. A property can be changed by selecting it and clicking DEFINE.** Preview → VIEW → DEFAULT ORIENTATION → OK **Refer Fig. 2.9.**

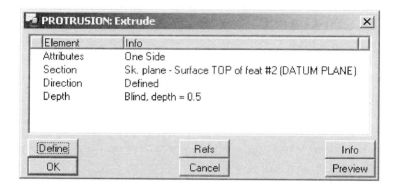

Fig. 2.8.

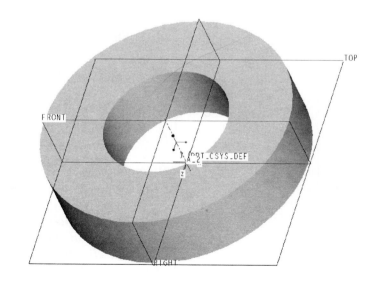

Fig. 2.9.

Goal	Step	Commands
Round the edges	16. Round the four edges of the bearing.	Create → Solid → Round → Simple → Done → Constant → **Creates a constant radius round between two surfaces.** Edge Chain → Done → Tangent Chain → Pick → **In the "tangent chain" option, the edges that are tangent to the selected edge will also be rounded.** *Select the four edges to be rounded →* **Refer Fig. 2.10.** Done → 0.025 → ✓ → OK **Refer Fig. 2.11.**

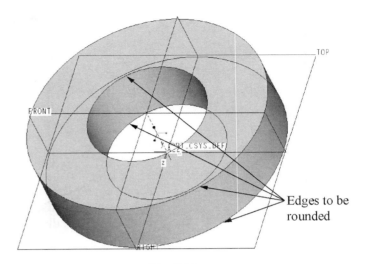

Edges to be rounded

Fig. 2.10.

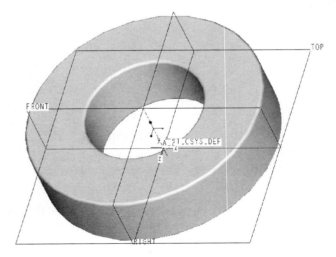

Fig. 2.11.

Goal	Step	Commands
View the model	17. Turn the datum planes off.	*Click on the following icons to switch off the datums, axis, datum points and default coordinate system.* **These icons help in turning the datum planes, axis, datum points and coordinate system on and off.** **Refer Fig. 2.12.** **Modifying the display may help in visualizing the model better. One of the four model display options can be selected by clicking on the corresponding icon:** **Wire-frame -** **Hidden line -** **No hidden line -** **Shaded -** **Fig. 2.13 shows the model in the four display types.**

Fig. 2.12.

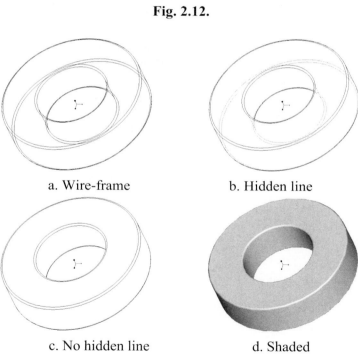

a. Wire-frame   b. Hidden line

c. No hidden line   d. Shaded

Fig. 2.13.

Goal	Step	Commands
Modify dimensions	18. Modify the dimensions.	▶ PART → Modify → Value → Pick → *Select the protrusion feature by clicking on the bearing in the graphics window or from the model tree →* *Select the 2.0 dimension →* <u>1.25</u> → ✓ → *Select the 1.0 dimension* → <u>0.6</u> → ✓ → Regenerate **The modifications take affect after the regeneration.** **Refer Fig. 2.14.**
Save the file and exit ProE	19. Save the file and exit ProE.	FILE → SAVE → <u>BEARING.PRT</u> → ✓ → FILE → EXIT → Yes

Fig. 2.14.

LESSON 3
BEARINGS

Learning Objectives:

- Model a part using different approaches.

- Evaluate different approaches for their ability to capture the design intent.

- Practice *Protrusion – Extrude* and *Round* features.

- Learn *Protrusion – Revolve* and *Hole* features.

Design Information:

The bearing can be modeled using a number of approaches. In this lesson, four distinct approaches are presented. Some of these approaches are more effective than others. The key factors that determine the effectiveness are:

- The ability of the model to capture the design intent.

- The flexibility to modify the model at a later time.

- The regeneration time.

- The required time to model the part.

The designer must consider alternative approaches, and then carefully choose an appropriate approach based on the task at hand.

Approach #1: Create the base cylinder by revolving a rectangular section.

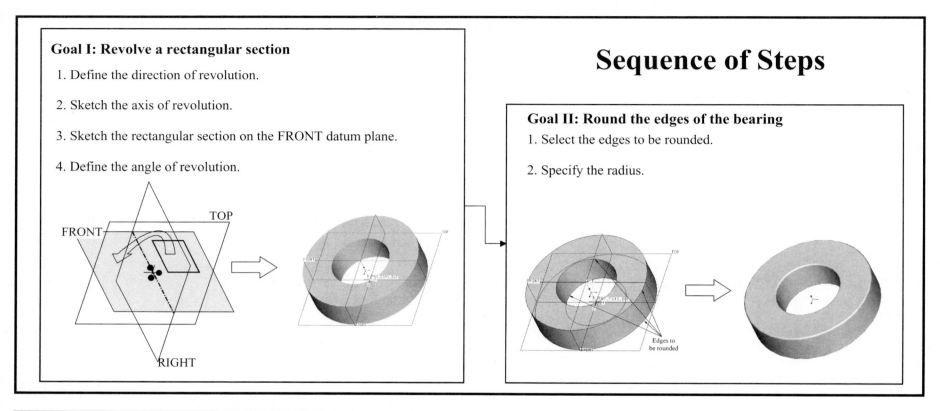

Goal I: Revolve a rectangular section

1. Define the direction of revolution.

2. Sketch the axis of revolution.

3. Sketch the rectangular section on the FRONT datum plane.

4. Define the angle of revolution.

Sequence of Steps

Goal II: Round the edges of the bearing

1. Select the edges to be rounded.

2. Specify the radius.

Goal	Step	Commands
Open a new file for the bearing part	1. Set up the working directory.	FILE → SET WORKING DIRECTORY → *Select the working directory* → OK
	2. Open a new file.	FILE → NEW → *Part* → *Solid* → bearing1 → OK
Create the base cylinder	3. Start "Protrusion – Revolve" feature.	Feature → Create → Solid → Protrusion → Revolve → Solid → Done

Revolve command sweeps the section around an axis of revolution. ProE uses the first centerline created in the sketcher as the axis of revolution. For the revolve feature, the entire section must lie to one side of the axis of revolution.

Lesson 3 – Bearings

Goal	Step	Commands
Create the base cylinder (Continued)	4. Define the direction of revolution.	One Side → Done **The "One side" option revolves the section by the specified angle in one direction. "Both sides" revolves the section in opposite directions.**
	5. Select the sketching plane.	**We are going to sketch the section on the FRONT datum plane.** Setup New → Plane → Pick *Select the FRONT datum plane by clicking on the word "FRONT"* **Refer Fig. 3.1.** **The red arrow points to the direction of feature creation.** Okay
	6. Orient the sketching plane.	Right → Plane → Pick → *Select the RIGHT datum plane by clicking on the word "RIGHT"*

Sketching plane

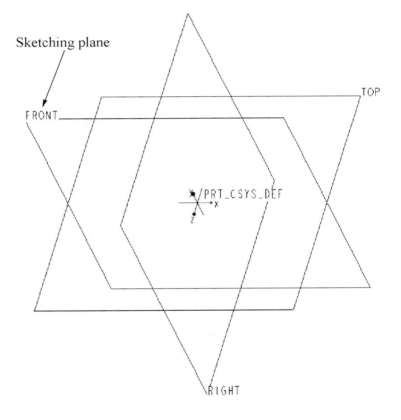

Fig. 3.1.

Mechanical Design Modeling Using ProEngineer   Page 3-3

Goal	Step	Commands
Create the base cylinder (Continued)	7. Sketch the axis of revolution on the RIGHT datum plane.	＼ ＼ ┊ → ┊ → *Pick points 1 and 2 on the RIGHT datum plane* **The cursor snaps onto the RIGHT datum if we move the cursor close to it.**
	8. Create a rectangular section.	▢ → *Pick points 3 and 4* **Refer Fig. 3.2.** **The first centerline should be the axis of revolution. For the revolve feature, the entire section must lie to one side of the axis of revolution.**
	9. Modify the dimensions.	↖ → *Double click the height dimension* → 0.5 → ***ENTER*** → *Double click the width dimension* → 0.5→ ***ENTER*** → *Double click the horizontal placement dimension* → 0.5→ ***ENTER*** **Refer Fig. 3.3.**
	10. Exit sketcher.	✔
	11. Define the angle of revolution.	360 → Done
	12. Accept the feature creation after previewing.	Preview→ VIEW → DEFAULT ORIENTATION → OK

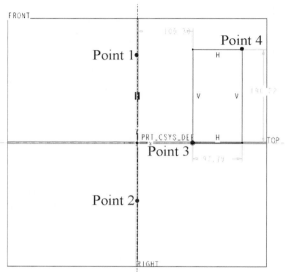

Fig. 3.2.

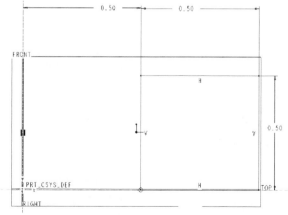

Fig. 3.3.

Goal	Step	Commands
Round the edges	13. Round the four edges of the cylinder.	Create → Solid → Round → Simple → Done → Constant → Edge Chain → Done → Tangent Chain → Pick → *Select the four edges to be rounded* → Done → 0.025 → ☑ → OK **Refer Fig. 3.4.**
View the model	14. View the model in the default view and turn the datum planes off.	VIEW → DEFAULT ORIENTATION → *Click on the following icons to switch off the datums, axis, datum points and default coordinate system.* **Refer Fig. 3.5.**
Save the file and exit ProE	15. Save the file and exit ProE.	FILE → SAVE → BEARING1.PRT → ☑ → FILE → EXIT

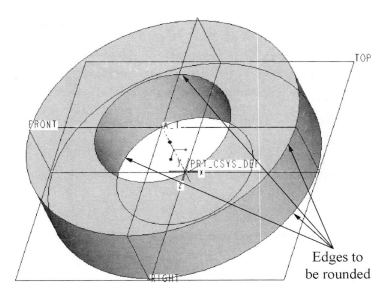

Edges to be rounded

Fig. 3.4.

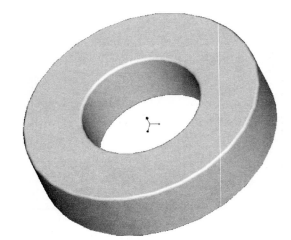

Fig. 3.5.

Approach #2: Create the base cylinder by revolving a line and defining its thickness.

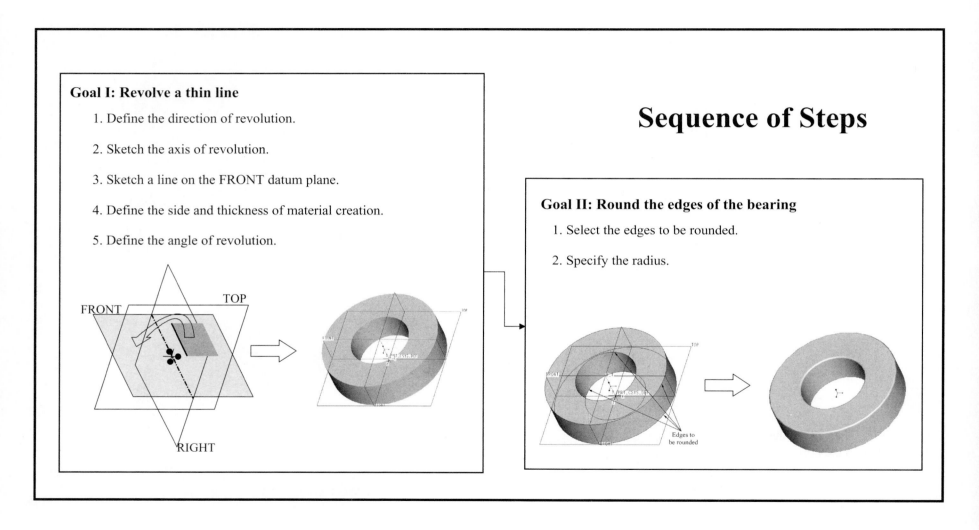

Goal I: Revolve a thin line

   1. Define the direction of revolution.

   2. Sketch the axis of revolution.

   3. Sketch a line on the FRONT datum plane.

   4. Define the side and thickness of material creation.

   5. Define the angle of revolution.

Sequence of Steps

Goal II: Round the edges of the bearing

   1. Select the edges to be rounded.

   2. Specify the radius.

Edges to
be rounded

Goal	Step	Commands
Open a new file for the bearing part	1. Set up the working directory.	FILE → SET WORKING DIRECTORY → *Select the working directory* → OK
	2. Open a new file.	FILE → NEW → *Part* → *Solid* → bearing2 → OK
Create the base cylinder	3. Start "Protrusion – Revolve" feature.	Feature → Create → Solid → Protrusion → Revolve → Thin ★ → Done
	4. Define the direction of revolution.	One Side → Done
	5. Select the sketching plane.	**We are going to sketch the section on the FRONT datum plane.** Setup New → Plane → Pick → *Select the FRONT datum plane* **Refer Fig. 3.6.** **The red arrow points to the direction of feature creation.** Okay
	6. Orient the sketching plane.	Right → Plane → Pick → *Select the RIGHT datum plane*

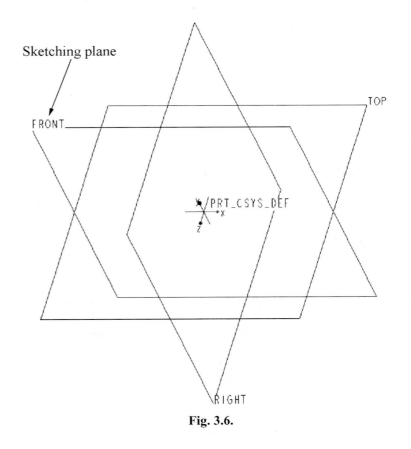

Sketching plane

Fig. 3.6.

Goal	Step	Commands
Create the base cylinder (Continued)	7. Sketch the axis of revolution on the RIGHT datum plane.	⟍ ⟍ ┊ → ┊ → *Pick points 1 and 2 on the RIGHT datum* **The cursor snaps onto the RIGHT datum plane if we move the cursor close to it.**
	8. Create a line.	┊ ⟍ ┊ → ⟍ → *Pick points 3 and 4 → Middle Mouse* (to discontinue line creation) **Refer Fig. 3.7.**
	9. Modify the dimensions.	⬉ *→ Double click the height dimension → 0.5 → ENTER → Double click the horizontal placement dimension → 0.5 → ENTER* **Refer Fig. 3.8.**
	10. Exit sketcher.	✔
	11. Define the side and the thickness for the entity.	Flip → Okay → 0.5 → ✅ **The arrow points to the direction of material addition.**
	12. Define the angle of revolution.	360 → Done
	13. Accept the feature creation after previewing.	Preview → VIEW → DEFAULT ORIENTATION → OK
	Readers are advised to round the edges. In case of problems, refer Step 13 in the previous approach.	

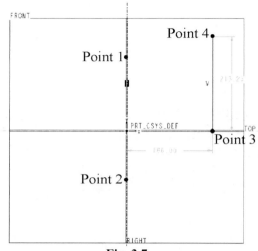

Fig. 3.7.

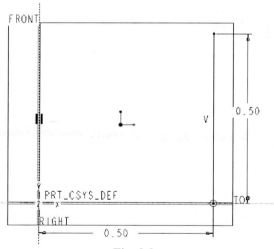

Fig. 3.8.

Approach #3: Create the base cylinder by extruding a circle and defining its thickness.

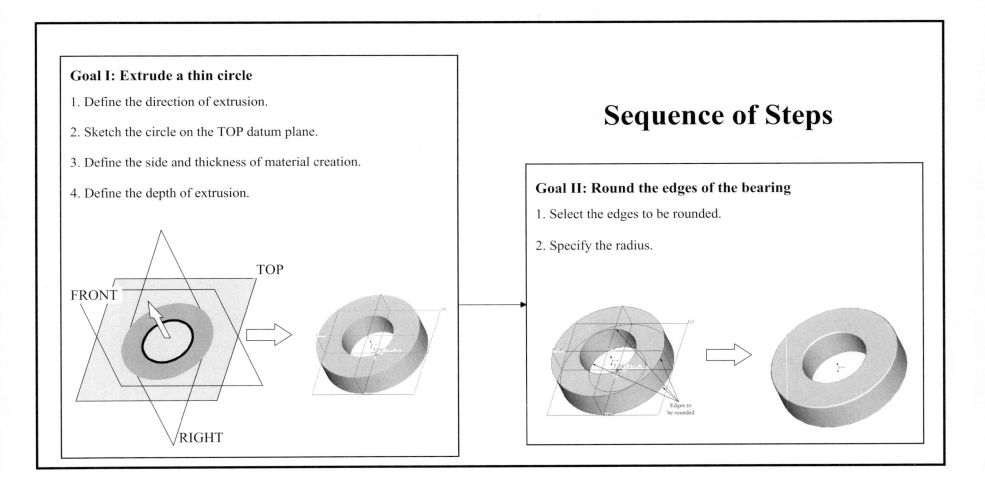

Goal I: Extrude a thin circle

1. Define the direction of extrusion.

2. Sketch the circle on the TOP datum plane.

3. Define the side and thickness of material creation.

4. Define the depth of extrusion.

TOP

FRONT

RIGHT

Sequence of Steps

Goal II: Round the edges of the bearing

1. Select the edges to be rounded.

2. Specify the radius.

Edges to be rounded

Goal	Step	Commands
Open a new file for the bearing part	1. Set up the working directory.	FILE → SET WORKING DIRECTORY → *Select the working directory* → OK
	2. Open a new file.	FILE → NEW → *Part* → *Solid* → bearing3 → OK
Create the base cylinder	3. Start "Protrusion – Extrude" feature.	Feature → Create → Solid → Protrusion → Extrude → Thin ★ → Done
	4. Define the direction of extrusion.	One Side → Done
	5. Select the sketching plane.	**We are going to sketch the section on the TOP datum plane.** Setup New → Plane → Pick *Select the TOP datum plane* **Refer Fig. 3.9.** **The red arrow points to the direction of feature creation.** Okay
	6. Orient the sketching plane.	Right → Plane → Pick → *Select the RIGHT datum plane*

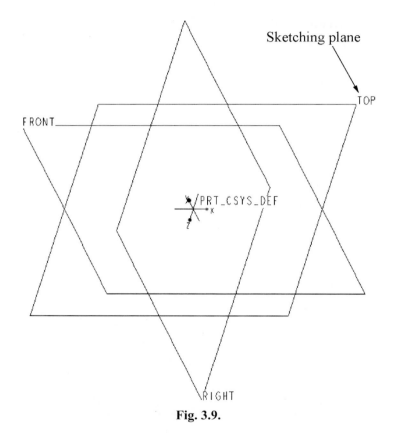

Fig. 3.9.

Goal	Step	Commands
Create the base cylinder (Continued)	7. Sketch a circle.	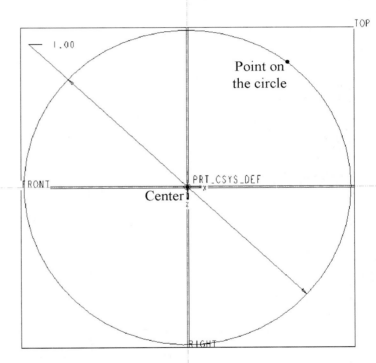 ⭘ → *Select the center of the circle at the intersection of front and right planes* **The cursor snaps onto the intersection.** *Select a point to define the outer edge of the circle* **Refer Fig. 3.10.**
	8. Modify the dimensions.	↖ → *Double click the diameter dimension* → 1 → ***ENTER*** **Refer Fig. 3.10.**
	9. Exit sketcher.	✔
	10. Define the side and the thickness for the entity.	Flip → Okay → 0.5 → ✅ **By flipping the arrow to the outside, the thickness will be added in the outward direction.**
	11. Define the extrusion depth.	Blind → Done → 0.5 → ✅
	12. Accept the feature creation after previewing.	Preview → VIEW → DEFAULT ORIENTATION → OK

Readers are advised to round the edges. In case of problems, refer Step 13 in the first approach.

Point on the circle

PRT_CSYS_DEF
Center

Fig. 3.10.

Approach #4: Create the base cylinder by extruding a circle. Then, place a hole at the center.

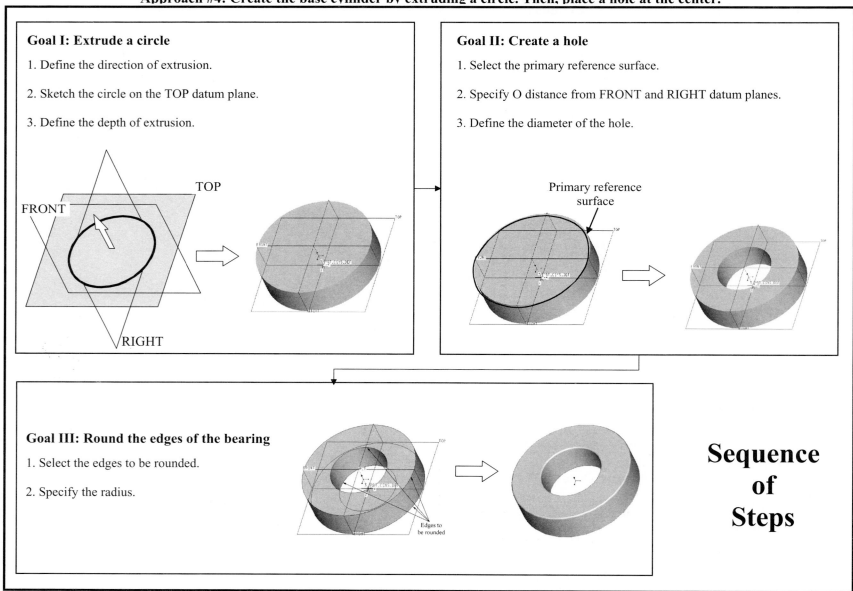

Goal I: Extrude a circle

1. Define the direction of extrusion.

2. Sketch the circle on the TOP datum plane.

3. Define the depth of extrusion.

Goal II: Create a hole

1. Select the primary reference surface.

2. Specify O distance from FRONT and RIGHT datum planes.

3. Define the diameter of the hole.

Goal III: Round the edges of the bearing

1. Select the edges to be rounded.

2. Specify the radius.

Sequence of Steps

Goal	Step	Commands
Open a new file for the bearing part	1. Set up the working directory.	FILE → SET WORKING DIRECTORY → *Select the working directory* → OK
	2. Open a new file.	FILE → NEW → *Part* → *Solid* → bearing4 → OK
Create the base cylinder	3. Start "Protrusion – Extrude" feature.	Feature → Create → Solid → Protrusion → Extrude → Solid → Done
	4. Define the direction of extrusion.	One Side → Done
	5. Select the sketching plane.	**We are going to sketch the section on the TOP datum plane.** Setup New → Plane → Pick *Select the TOP datum plane* **Refer Fig. 3.11.** **The red arrow points to the direction of feature creation.** Okay
	6. Orient the sketching plane.	Right → Plane → Pick → *Select the RIGHT datum plane*

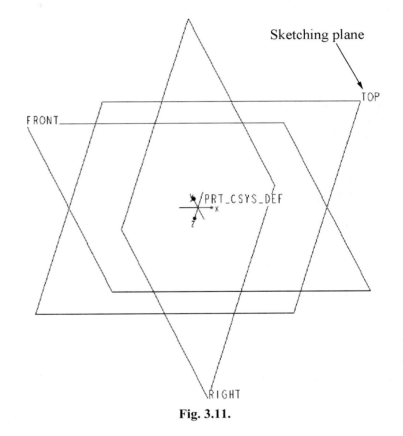

Sketching plane

TOP

FRONT

PRT_CSYS_DEF

RIGHT

Fig. 3.11.

Goal	Step	Commands
Create the base cylinder (Continued)	7. Sketch a circle.	⭕ → *Select the center of the circle as the intersection of front and right planes* **The cursor snaps onto the intersection.** *Select a point to define the outer edge of the circle* **Refer Fig. 3.12.**
	8. Modify the dimensions.	↖ → *Double click the diameter dimension* → 2 → ***ENTER*** **Refer Fig. 3.12.**
	9. Exit sketcher.	✔
	10. Define the extrusion depth.	Blind → Done → 0.5 → ✔
	11. Accept the feature creation after previewing.	Preview → VIEW → DEFAULT ORIENTATION → OK **Refer Fig. 3.13.**

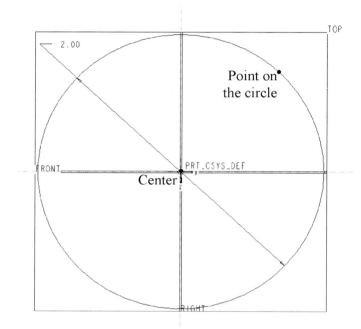

Fig. 3.12.

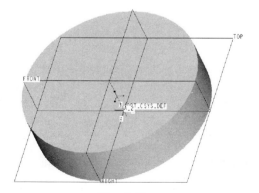

Fig. 3.13.

Goal	Step	Commands
Create a hole	12. Create a hole at the center.	Create→ Solid→ Hole → **ProE opens the "hole" window. Refer Fig. 3.14.** (Hole Type) Straight → (Diameter) 1.0 → (Depth One) *Thru All* → (Primary Reference) → Query Selec → *Click the top circular surface* → **ProE opens the Query Bin window. If we click on an item in the Query Bin window, ProE highlights it in the graphics window.** *Click each item in the Query Bin window until the top circular surface is highlighted* → ACCEPT → (Linear Reference) → *Select the FRONT datum plane from the model tree* → (Distance) 0 → (Linear Reference) → *Select the RIGHT datum plane from the model tree* → (Distance) 0 →

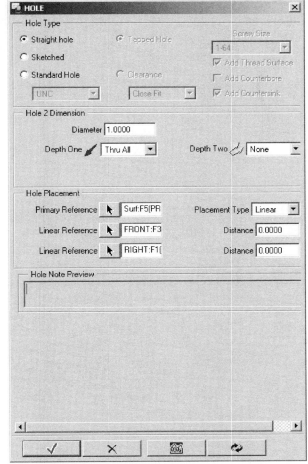

Fig. 3.14.

Primary reference refers to the plane on which the hole is placed. The distances from the linear references determine the location of the hole.

Readers are advised to round the edges. In case of problems, refer Step 13 in the first approach.

Query Select is a useful command for selecting a specific entity in a cluster of entities or a hidden entity. The RIGHT mouse initiates the query select mode. After the initiation, click LEFT mouse in the close vicinity of the entity. ProE opens Query Bin and shows possible entities that lie in the vicinity. The RIGHT mouse or UP and DOWN arrow keys can be used to select the appropriate entity. Then, click on ACCEPT or MIDDLE mouse button. Fig. 3.15 shows the mouse button functions. Query select is also useful in selecting hidden geometry.

Left mouse	Middle mouse	Right mouse
Pick	Accept	Start query select
		Next entity

Fig. 3.15.

Discussion

Criteria	Approaches			
	Approach #1: Create the base cylinder by revolving a rectangular section.	*Approach #2: Create the base cylinder by revolving a line and defining its thickness.*	*Approach #3: Create the base cylinder by extruding a circle and defining its thickness.*	*Approach #4: Create the base cylinder by extruding a circle. Then, place a hole at the center.*
Ability to capture the design intent.	**GOOD** The approach captures the design intent by clearly indicating the two critical dimensions (outer and inner diameters).	**NOT ACCEPTABLE** Even though the final part looks identical, this approach does not capture the design intent. In this approach, we specify the inner diameter and the thickness.		**GOOD** The approach captures the design intent by clearly indicating the two critical dimensions (outer and inner diameters).
Flexibility in modifying the model at a later time	**GOOD** It is easy to modify the inner and outer diameters.	**ACCEPTABLE** In this approach, we have to modify the inner diameter and thickness.		**EXCELLENT** This approach provides the most flexibility. It is very easy to modify dimensions. Also, if necessary, it is possible to suppress the hole feature.

Learning Objectives:

- Practice **Protrusion – Revolve** and **Round** features.

- Create simple sketches.

- Learn **Chamfer** and **Cut – Revolve** features.

- Control the model display.

Design Information:

Bushings are the simplest and the cheapest form of bearings. They are used to support both rotating and translating components while using a small amount of radial space. As the bushing is in physical contact with the rotating shaft, wear is a major problem. Therefore, replacement cost should be included in the total life cycle cost considerations.

A tight fit between the bushing and the housing, and a running fit between the shaft and the bushing are specified. The length to diameter ratio is a key parameter that determines the performance of the bushing. Lubrication becomes a problem when the L/D ratio is less than one whereas alignment becomes a problem when the L/D ratio is greater than four.

Sequence of Steps

Goal I: Revolve a rectangular section

1. Define the direction of revolution.

2. Sketch the axis of revolution.

3. Sketch the section on the FRONT datum plane.

4. Define the angle of revolution.

Goal II: Round the edges

1. Select the edges to be rounded.

2. Specify the radius.

Goal III: Chamfer the lower edge

1. Select the edge to be rounded.

2. Specify the radius.

Goal IV: Create a revolve cut

1. Define the direction of revolution.

2. Sketch the axis of revolution.

3. Sketch the section on the FRONT datum plane.

4. Define the direction of material removal.

5. Define the angle of revolution.

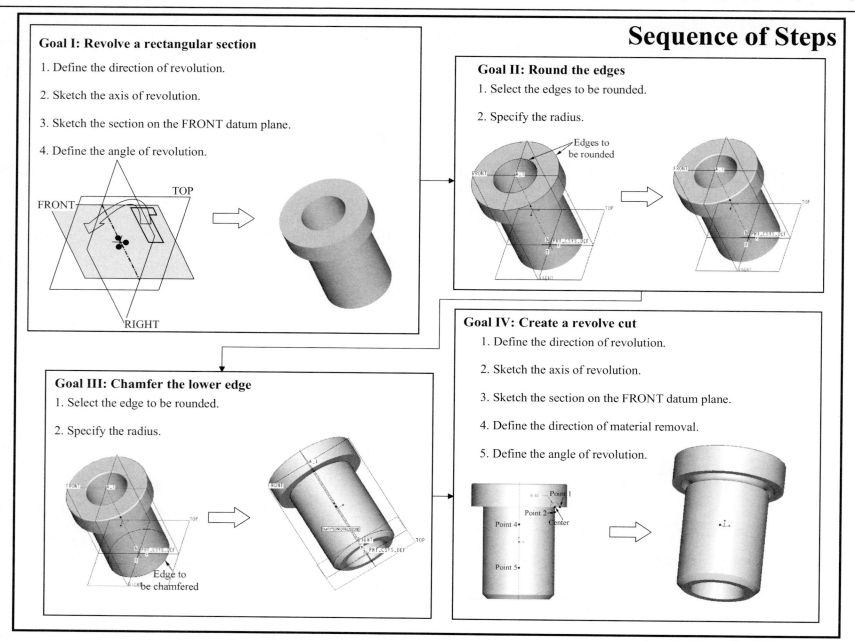

Goal	Step	Commands
Open a new file for the bushing part	1. Set up the working directory.	FILE → SET WORKING DIRECTORY → *Select the working directory* → OK
	2. Open a new file.	FILE → NEW → *Part* → *Solid* → bushing → OK
Create the base feature	3. Start "Revolve – Solid" feature.	Feature → Create → Solid → Protrusion → Revolve → Solid → Done
	4. Define the direction of revolution.	One side → Done
	5. Select the sketching plane.	Setup New → Plane → Pick → *Select the FRONT datum plane* → Okay
	6. Orient the sketching plane.	Right → Plane → Pick → *Select the RIGHT datum plane*
	7. Sketch the axis of revolution on the RIGHT datum plane.	[icons] → [icon] → *Pick points 1 and 2 on the RIGHT datum plane*
	8. Sketch the section.	[icons] → [icon] → *Pick points 3, 4, 5, 6, 7, 8 and 3* → *Middle Mouse* (to discontinue line creation) **Refer Fig. 4.1.**

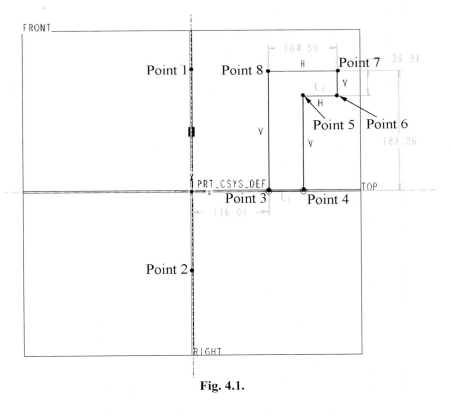

Fig. 4.1.

Goal	Step	Commands
Create the base feature (Continued)	9. Create new dimensions	⊢↔⊣ → *Select lines 1 and 2 → Middle Mouse at the point where the vertical dimension must be placed →* **Refer Fig. 4.2.** **The position of the cursor determines the type of dimension placed (horizontal, or vertical).** *Select lines 1 and 3 → Middle Mouse to place the vertical dimension →* *Select line 4, centerline and again on line 4 → Middle Mouse to place the diameter dimension →* **Note that ProE places the diameter dimension.** *Select line 5, centerline and again on line 5 → Middle Mouse to place the diameter dimension → Select line 6, centerline and again on line 6 → Middle Mouse to place the diameter dimension*

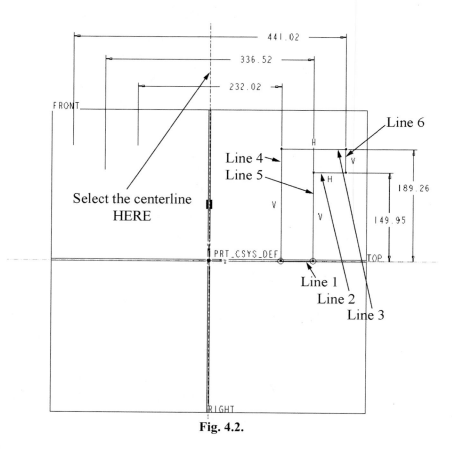

Fig. 4.2.

When an entity is created, ProE automatically places weak dimensions (shown in gray color). ProE erases these dimensions depending on the constraints. We can create strong dimensions (shown in yellow) by either specifying values to the weak dimensions or by creating new dimensions. ProE will not erase the strong dimensions.

Goal	Step	Commands
Create the base feature (Continued)	10. Modify the dimensions.	★ **Modify the smaller dimensions first, and then the larger dimensions.** ➤ → *Double click each dimension and enter the corresponding value* **Refer Fig. 4.3.**
	11. Exit sketcher.	✔
	12. Define the angle of revolution.	360 → Done
	13. Accept the feature creation after previewing.	Preview → VIEW → DEFAULT ORIENTATION → OK **Refer Fig. 4.4.**
Round the edges	14. Round the two outside ends of the bushing.	Create → Solid → Round → Simple → Done → Constant → Edge Chain → Done → Tangent Chain → Pick → *Select the two edges to be rounded* → **Refer Fig. 4.4.** Done → 0.016 → ✔ → OK **Refer Fig. 4.5.**

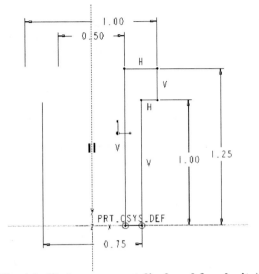

Fig. 4.3. (Datums are not displayed for clarity)

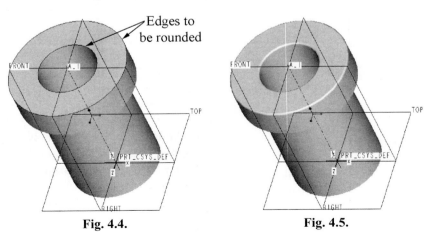

Edges to be rounded

Fig. 4.4. Fig. 4.5.

Goal	Step	Commands
Chamfer the bottom outer edge	15. Chamfer the outer edge.	Create → Solid → Chamfer → Edge → 45 x d → 0.05 → ✔ → Pick → *Select the bottom edge* → **Refer Fig. 4.6.** Done Sel → Done Refs → OK **Refer Fig. 4.7.**
Create an undercut	16. Start "Cut – Revolve" feature.	Create → Solid → Cut → Revolve → Solid → Done
	17. Define the direction of revolution.	One Side → Done
	18. Select the previous sketching plane.	Use Prev → Okay
	19. Add new references.	**If the References window is not visible, then select SKETCH → REFERENCES command.** *Select the vertical and horizontal edges* **Refer Figs. 4.8 and 4.9.**

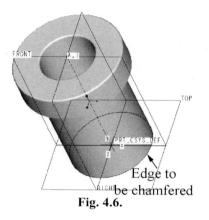

Edge to be chamfered

Fig. 4.6.

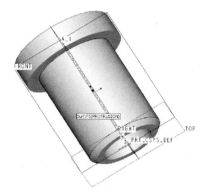

Fig. 4.7.

Fig. 4.8.

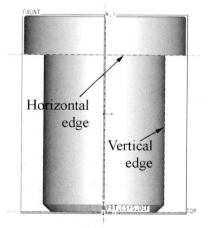

Horizontal edge

Vertical edge

Fig. 4.9.

Goal	Step	Commands
Create an undercut (Continued)	20. Sketch an arc.	⌐ | ▸ ⌐ ⌐ ↗ ⌐ ⌐ → ⌐ → *Select the center →* **Refer Fig. 4.10.** *Pick points 1 and 2* **Refer Fig. 4.10.**
	21. Modify the arc radius.	↖ → *Double click radius →* <u>0.05</u> → ***ENTER***
	22. Create the axis of revolution.	↘ | ▸ ↘ ┆ → ┆ → *Pick points 4 and 5 on the RIGHT datum plane* **Refer Fig. 4.10.**
	23. Exit sketcher.	✔
	24. Define the direction of material removal.	Okay **The arrow points to the direction of the material removal.**
	25. Define the angle of revolution.	360 → Done
	26. Accept the feature creation after previewing.	Preview → VIEW → DEFAULT ORIENTATION → OK → *Rotate the model using* ***CNTRL*** *key + Left Mouse* **Refer Fig. 4.11.**

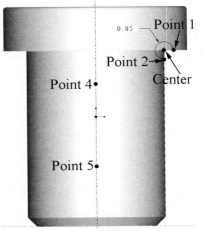

Fig. 4.10.

Fig. 4.11.

Goal	Step	Commands
View the model	27. Define the model color.	VIEW → MODEL SETUP → COLOR & APPEARANCES → **ProE opens "Appearances" window.** ADD → **ProE opens "Appearance Editor" window.** **Refer Fig. 4.12.** *Click on the white area* → **ProE opens the "Color Editor" window.** **Refer Fig. 4.13.** *Click on Color wheel* → **ProE opens the color wheel.** *Select a suitable color* → CLOSE → OK → SET → Close

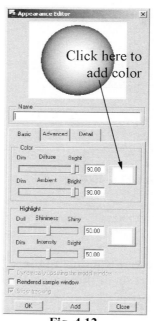

Fig. 4.12.

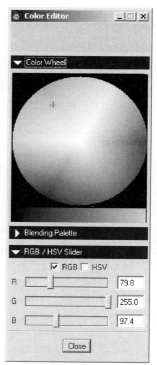

Fig. 4.13.

Goal	Step	Commands
View the model (Continued)	28. View the model in the default view. Turn off the datum planes.	VIEW → DEFAULT ORIENTATION → *Click on the following icons to switch off the datums, axis, datum points and default coordinate system.* □ ⌀ ⅏ ⅏ → UTILITIES → ENVIRONMENT → **ProE opens the "Environment" window.** *Uncheck Spin Center* →. OK **Refer Fig. 4.14.**
Save the file and exit ProE	29. Save the file and exit ProE.	FILE → SAVE → BUSHING.PRT → ✓ → FILE → EXIT → Yes

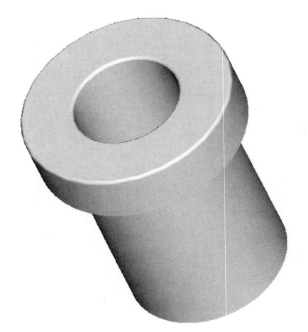

Fig. 4.14.

LESSON 5
RETAINING RING

Learning Objectives:

- Create a part with complex geometry using simple features.

- Practice *Protrusion – Extrude*, *Cut – Extrude*, *Round* and *Hole* features.

- Learn *Mirror* command.

Design Information:

Retaining rings are used to:

- Locate and secure machine components on a shaft by providing an accurate locating shoulder.

- Secure components in the housing.

They are placed in locating grooves machined on the shafts or in the bores. Depending on the placement of the groove, they are classified as external or internal rings.

Retaining rings make the products easy to assemble during production and disassemble for service. They eliminate expensive machining operations that are typically required for threaded connections. They also eliminate the shaft extensions that are required for the threaded parts.

Sequence of Steps

Goal I: Create the base feature

1. Define the direction of extrusion.

2. Sketch the section on the TOP datum plane.

3. Define the depth of extrusion.

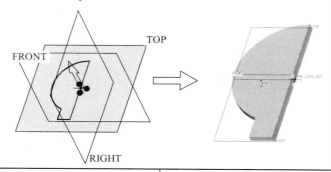

Goal II: Mirror geometry

1. Mirror the geometry about the RIGHT datum plane.

Goal III: Create a hole

1. Define the diameter of the hole.

2. Select the primary reference surface.

3. Specify the distance from FRONT and RIGHT datum planes.

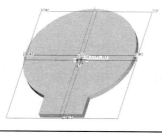

Goal IV: Create a small hole

1. Define the diameter of the hole.

2. Select the primary reference surface.

3. Specify the distance from FRONT and RIGHT datum planes.

Goal V: Mirror feature

1. Mirror the hole about the RIGHT datum plane.

Goal VI: Cut a vertical slot

1. Sketch the section on the TOP datum plane.

2. Define the direction and the depth of cut.

Goal VII: Round the edges

1. Select the surfaces between which the round feature must be created.

2. Specify the radius.

Goal VIII: Round the edges

1. Select the edges to be rounded.

2. Specify the radius.

Goal	Step	Commands
Open a new file for the retaining ring part	1. Set up the working directory.	FILE → SET WORKING DIRECTORY → *Select the working directory* → OK
	2. Open a new file.	FILE → NEW → *Part* → *Solid* → ring → OK
Create the base feature	3. Start "Extrude – Solid" feature.	Feature → Create → Solid → Protrusion → Extrude → Solid → Done
	4. Define the direction of extrusion.	One side → Done
	5. Select the sketching plane.	Setup New → Plane → Pick → *Select the TOP datum plane* → Okay
	6. Orient the sketching plane.	Right → Plane → Pick → *Select the RIGHT datum plane*
	7. Create an arc.	⌐ ▸ ⌐ ⌇ ⌒ ⌒ → ⌒ → *Pick center on RIGHT datum* → *Pick points 1 and 2* **Refer Fig. 5.1.**
	8. Create three lines.	＼ → *Pick points 2, 3, 4, and 1* → *Middle Mouse* (to discontinue line creation) **Refer Fig. 5.2.**

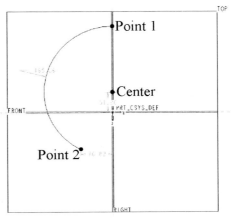

Fig. 5.1.

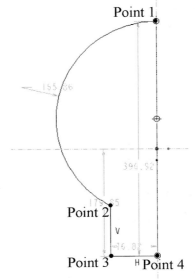

Fig. 5.2.

Goal	Step	Commands
Create the base feature (Continued)	9. Add dimensions.	|↔| → *Select the arc center and then the PRT_CSYS_DEF* → *Middle Mouse at the point where the vertical dimension must be placed* → *Select line 1* → *Middle Mouse to place the vertical dimension* → *Select line 2* → *Middle Mouse to place the horizontal dimension* **Refer Fig. 5.3.**
	10. Modify dimensions.	★**Modify the linear dimensions first and then, the radius.** ↖ → *Double click on each dimension and enter the corresponding value* **Refer Fig. 5.3.**
	11. Exit sketcher.	✔
	12. Define the extrusion depth.	Blind → Done → <u>0.025</u> → ✓
	13. Accept the feature creation after previewing.	Preview → VIEW → DEFAULT ORIENTATION → OK **Refer Fig. 5.4.**
Mirror the extrude feature	14. Mirror the geometry.	Mirror Geom → Plane → Pick → *Select the RIGHT datum plane in the model tree* **Refer Fig. 5.5.**

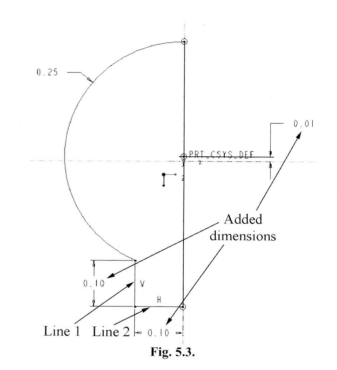

Fig. 5.3.

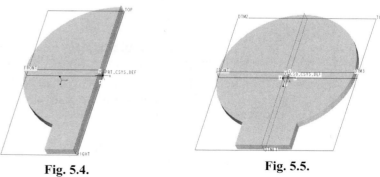

Fig. 5.4.   Fig. 5.5.

Goal	Step	Commands
Create a large hole	15. Create a hole at the center.	Create→ Solid→ Hole → **ProE opens the "hole" window.** **Refer Fig. 5.6.** (Hole Type) Straight → (Diameter) 0.4 → (Depth One) *Thru All* → (Primary Reference) → *Select the top surface of the base feature* → (Linear Reference) → *Select the FRONT datum plane from the model tree* → (Distance) 0 → (Linear Reference) → *Select the RIGHT datum plane from the model tree* → (Distance) 0 → **Refer Fig. 5.6.**

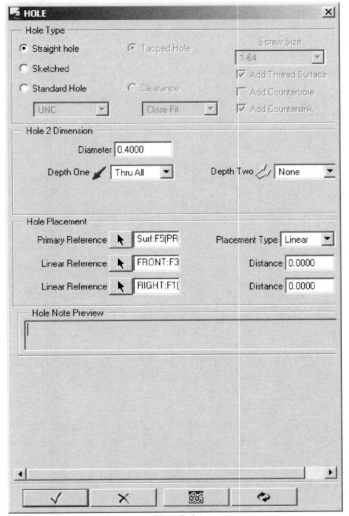

Fig. 5.6.

Goal	Step	Commands
Create a small hole	16. Create a small hole.	Create→ Solid→ Hole → **ProE opens "hole" window.** **Refer Fig. 5.7.** (Hole Type) Straight → (Diameter) 0.05 → (Depth One) *Thru All* → (Primary Reference) ↖ → *Select the top surface of the base feature* → (Linear Reference) ↖ → *Select the FRONT datum plane from the model tree* → (Distance) 0.25 → (Linear Reference) ↖ → *Select the RIGHT datum plane from the model tree* → (Distance) 0.05 → ✓ **Refer Fig. 5.7.**
Mirror the small hole	17. Start "Copy – Mirror" command.	Copy → Mirror → Select → Dependent → Done
	18. Pick the feature to be mirrored.	Select → Pick → *Select the small hole (last feature) from the model tree* → Done Sel → Done
	19. Pick the RIGHT datum plane.	Plane → Pick → *Select the RIGHT datum plane* **Refer Fig. 5.8.**

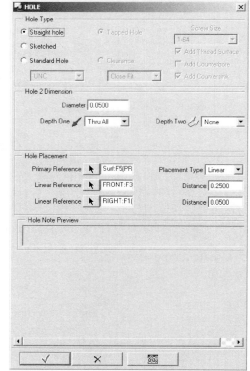

Fig. 5.7.

When the dependent option is selected, modifying the original hole also modifies the mirrored hole.

Fig. 5.8.

Goal	Step	Commands
Create the vertical cut	20. Start "Extrude – Cut" feature.	Create → Solid → Cut → Extrude → Solid → Done
	21. Accept One Side as the extrusion direction.	One side → Done
	22. Use the previous sketcher plane.	Use Prev **The "Use Prev" command uses the previous sketching plane.**
	23. Define the direction of cut.	**The default direction of a cut is away from you.** Flip ★ → Okay **At this point, the arrow should point towards you.**
	24. Add new references.	**If the references window is not visible, then select SKETCH → REFERENCES command.** **Refer Fig. 5.9.** *Select inner circle → Select the edge shown in Fig. 5.10.* **Refer Fig. 5.10.**
	25. Sketch the cut section.	▢ → *Pick points 1 and 2* **Refer Fig. 5.10.**

Fig. 5.9.

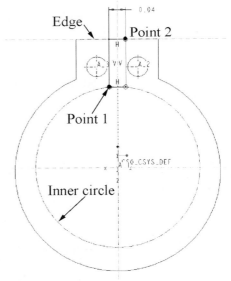

Fig. 5.10.

Goal	Step	Commands
Create the vertical cut (Continued)	26. Modify the width dimension.	↖ → *Double click on the width dimension* → 0.04 → ***ENTER***
	27. Exit sketcher.	✔
	28. Define the material removal direction.	Okay
	29. Define the depth of cut	Thru All → Done
	30. Accept the feature creation after previewing.	Preview → VIEW → DEFAULT ORIENTATION → OK **Refer Fig. 5.11.**
Round the edges	31. Round the two inner edges.	Create → Solid → Round → Simple → Done → Constant → Surf-Surf → Done → Query select surfaces 1 and 2 **Refer Fig. 5.11.** Enter → New Value → 0.1 → ✔ → OK Create → Solid → Round → Simple → Done → Constant → Surf-Surf → Done → Query select surfaces 3 and 4 **Refer Fig. 5.11.** Enter → New Value → 0.1 → ✔ → OK **Refer Fig. 5.12.**

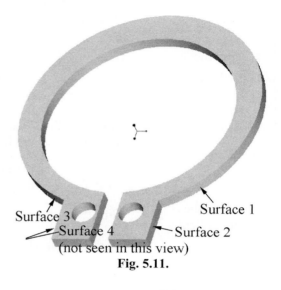

Surface 3
Surface 4
(not seen in this view)
Surface 1
Surface 2

Fig. 5.11.

Fig. 5.12.

Goal	Step	Commands
Round the edges (Continued)	32. Round the two outer edges.	Create → Solid → Round → Simple → Done → Constant → Edge Chain → Done → Tangent Chain → Pick → *Select the two sharp edges* → **Refer Fig. 5.13.** Done → Enter → New Value → 0.05 → ✓ → OK **Refer Fig. 5.14.**
View the model	33. View the model in the default view. Turn off the datum planes.	VIEW → DEFAULT ORIENTATION → *Click on the following icons to switch off the datums, axis, datum points and default coordinate system.* ▢▢▢▢ → UTILITIES → ENVIRONMENT → **ProE opens the "Environment" window.** *Uncheck Spin Center* → OK **Refer Fig. 5.14.**
Save the file and exit ProE	34. Save the file and exit ProE.	FILE → SAVE → RING.PRT → ✓ → FILE → EXIT → Yes

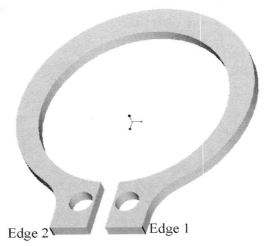

Edge 2   Edge 1

Fig. 5.13.

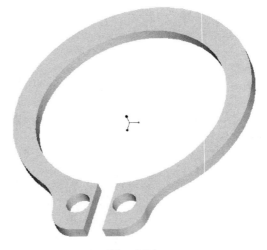

Fig. 5.14.

Learning Objectives:

- Practice *Protrusion – Revolve, Cut – Extrude, Hole, Chamfer* and *Round* features.

- Practice *Mirror* command in the sketcher.

- Learn *Make Datum* and *Pattern* commands.

- Learn the use of *Layers*.

Design Information:

Shafts are used for transmitting rotational energy (rotary motion and torque) from one location to another. They are stepped so that machine components such as cams and bearings can be mounted easily. The steps cause stress concentration and result in the premature failure of the shaft. The steps are rounded to reduce stress concentration. The radius of the round is a key parameter as an improper value prevents the proper assembly of the bearings.

Keys transmit the torque between the shaft and the machine element. The shape of the keyway depends on the type of the key and also, the manufacturing process (edge- or end-milling). Setscrews are often used to hold the keys in place.

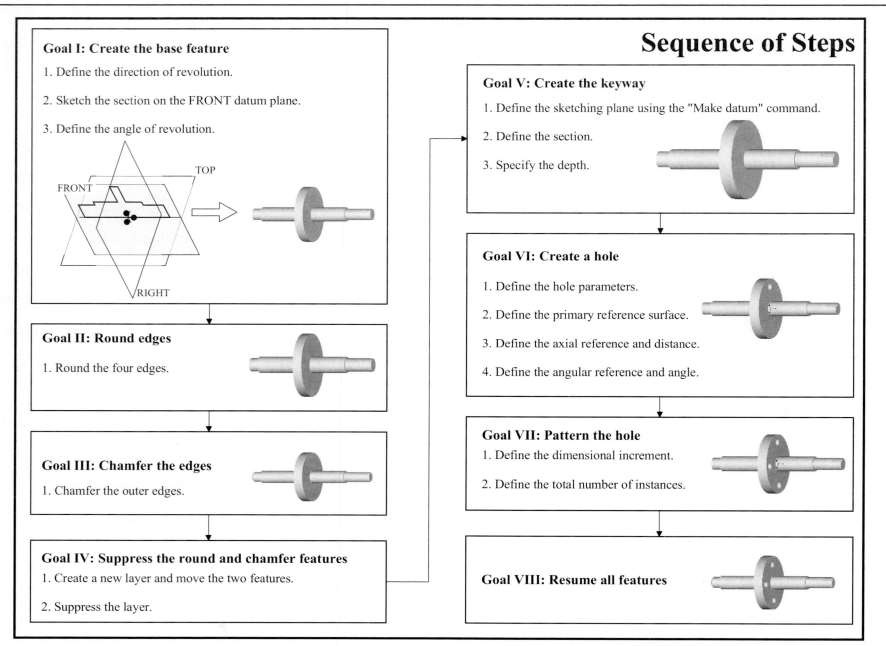

Sequence of Steps

Goal I: Create the base feature

1. Define the direction of revolution.

2. Sketch the section on the FRONT datum plane.

3. Define the angle of revolution.

TOP

FRONT

RIGHT

Goal II: Round edges

1. Round the four edges.

Goal III: Chamfer the edges

1. Chamfer the outer edges.

Goal IV: Suppress the round and chamfer features

1. Create a new layer and move the two features.

2. Suppress the layer.

Goal V: Create the keyway

1. Define the sketching plane using the "Make datum" command.

2. Define the section.

3. Specify the depth.

Goal VI: Create a hole

1. Define the hole parameters.

2. Define the primary reference surface.

3. Define the axial reference and distance.

4. Define the angular reference and angle.

Goal VII: Pattern the hole

1. Define the dimensional increment.

2. Define the total number of instances.

Goal VIII: Resume all features

Goal	Step	Commands
Open a new file for the shaft part	1. Set up the working directory.	FILE → SET WORKING DIRECTORY → *Select the working directory* → OK
	2. Open a new file.	FILE → NEW → *Part* → *Solid* → shaft → OK
Create the base feature	3. Start "Protrusion – Revolve" feature.	Feature → Create → Solid → Protrusion → Revolve → Solid → Done
	4. Define the direction of revolution.	One side → Done
	5. Select the sketching plane.	Setup New → Plane → Pick → *Select the FRONT datum plane* → Okay
	6. Orient the sketching plane.	Right → Plane → Pick → *Select the RIGHT datum plane*
	7. Sketch the axis of revolution on the RIGHT datum plane.	＼ ｜ ＼ ┆ → ┆ → *Pick points 1 and 2 on the TOP datum*
	8. Sketch the section.	┆ ＼ ┆ → ＼ → *Pick points 3, 4, 5, 6, 7, 8, 9, 10, 11, 12, 13, 14 and 3* → *Middle Mouse* **Refer Fig. 6.1.**
	9. Align points 7 and 10, and 5 and 12.	SKECTCH → CONSTRAIN or  **Refer Fig. 6.2.** ↔ → *Select points 7 and 10* → *Select points 5 and 12*

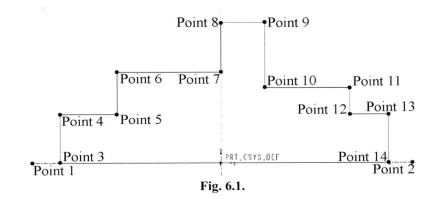

Fig. 6.1.

↕ Makes line or vertices vertical	↔ Makes line or vertices horizontal	⊥ Makes two entities perpendicular
⅋ Makes two entities tangential	↘ Places a point in the middle of the line	⊙ Creates coincident points or collinear constraint
⇥⇤ Makes two points or vertices symmetric about the centerline	= Makes the lengths, radii or curvatures equal	// Makes two lines parallel

Fig. 6.2.

Goal	Step	Commands
Create the base feature (Continued)	10. Create all horizontal dimensions from the RIGHT datum plane.	⊬⊢ → *Click on the RIGHT datum and line 1* → *Middle Mouse* → *Click on the RIGHT datum and line 2* → *Middle Mouse*→ *Click on the RIGHT datum and line 3* → *Middle Mouse*→ *Click on the RIGHT datum and line 4* → *Middle Mouse* → *Click on the RIGHT datum and line 5* → *Middle Mouse* **Refer Fig. 6.3.**
	11. Create new diameter dimensions.	⊬⊢ → *Select line 6, centerline and again line 6* → *Middle Mouse to place the diameter dimension* → *Select line 7, centerline and again line 7* → *Middle Mouse* → *Select line 8, centerline and again line 8* → *Middle Mouse* **Refer Fig. 6.3.**
	12. Modify the dimensions.	★ **First modify the smaller dimensions.** ↖ → *Double click each dimension and enter the corresponding value* **Refer Fig. 6.4.**

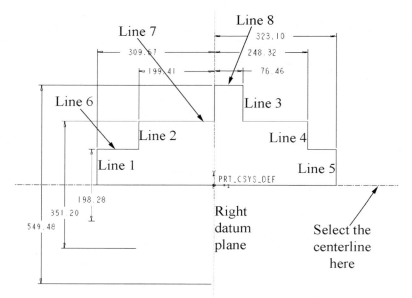

Fig. 6.3.

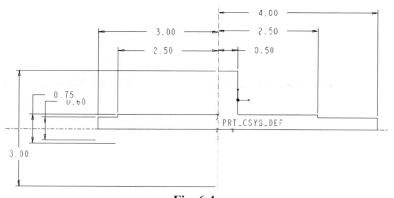

Fig. 6.4.

Goal	Step	Commands
Create the base feature (Continued)	13. Exit sketcher.	✔
	14. Define the angle of revolution.	360 → Done
	15. Accept the feature creation after previewing.	Preview → VIEW → DEFAULT ORIENTATION → OK **Refer Fig. 6.5.**
Round the edges	16. Round the four edges.	Create → Solid → Round → Simple → Done → Constant → Edge Chain → Done → Tangent Chain → Pick → *Query select the four edges to be rounded* → **Refer Fig. 6.6.** Done → 0.02 → ✔ → OK **Refer Fig. 6.7.**

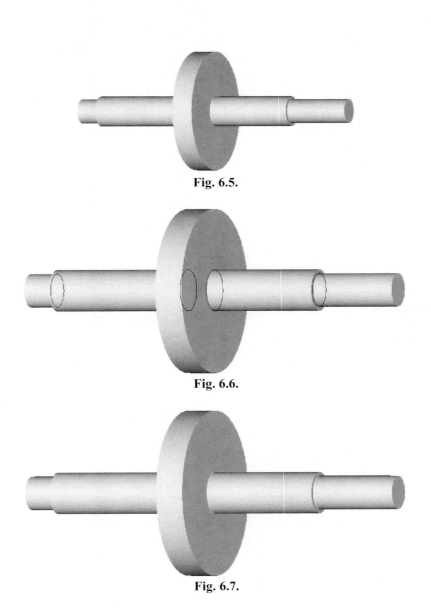

Fig. 6.5.

Fig. 6.6.

Fig. 6.7.

Goal	Step	Commands
Chamfer the outer edges	17. Chamfer the outer edges.	Create → Solid → Chamfer → Solid → Edge → 45 x d → 0.05 → ✓ → Pick → *Select the two edges* → **Refer Fig. 6.8.** Done Sel → Done Refs → OK **Refer Fig. 6.9.**
Move the round and chamfer features into a new layer	18. Create a new layer.	VIEW → LAYERS → **ProE opens a new window.** **Refer Fig. 6.10.** ⬚ → (Name) Rounds_Chamfers → ADD → Cancel (to close the "new layer" window)
	19. Move the round and chamfer features to the new layer.	*Select Rounds_Chamfers* → ➕ → Feature → Select → Pick → *Select the round and chamfer features from the model tree* → Done Sel → Done/Return → Done/Return

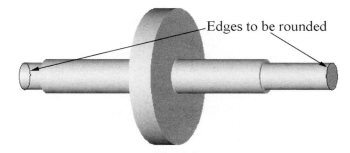

Edges to be rounded

Fig. 6.8.

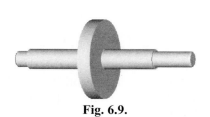

Fig. 6.9.

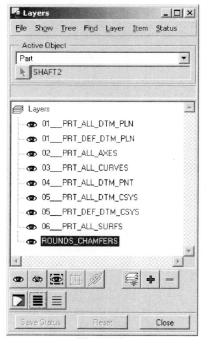

Fig. 6.10.

Goal	Step	Commands
Suppress the rounds and the chamfers	20. Suppress the rounds_chamfers layer.	**When a layer is set to blank, the solid geometry will not be affected. However, the geometric features can be suppressed by suppressing the layer.** CLOSE → Suppress → Normal → Layer → *Check Rounds_Chamfers* → Done Sel → Done **Now, the round and chamfer features should disappear.**
Create the keyway	21. Start "Cut – Extrude" feature.	Create → Solid → Cut → Extrude → Solid → Done → One Side → Done
	22. Set up the sketching plane.	Setup New → Make Datum → Offset → Plane → Plane → Coord Sys → Pick → *Select the TOP datum plane* → Enter Value → 0.3 → ✓ → Done → Okay → Right → *Select the RIGHT datum plane*
	23. Add additional references.	**If references window is not visible, then activate it using SKETCH → REFERENCES command.** *Add the right edge of the shaft to the existing references.* **Refer Fig. 6.11.**

Layers are useful for organizing the model. Several features can be deleted, reordered, suppressed or resumed using layers. In ProE, a feature can be assigned to multiple layers. The display status of a layer can be set to:

- Show – Selected layers are displayed.

- Blank – Selected layers are blanked. Only datum features, feature axes, cosmetic features and quilts can be blanked.

- Isolate – Only the selected layers are displayed. All other layers are blanked.

- Hidden – This option is available in the assembly mode only. Components in the hidden layers blanked in accordance with the Environment for the hidden-line display.

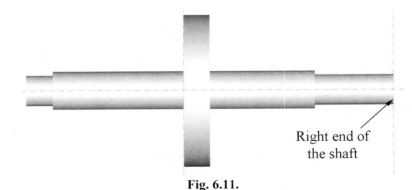

Right end of the shaft

Fig. 6.11.

Goal	Step	Commands
Create the keyway (Continued)	24. Sketch the key section.	$\setminus$ → *Pick points 1, 2 and 3* → *Middle Mouse* → $\frown$ → *Pick points 3 and 4 (Point 4 should be on the TOP datum)* → $\setminus$ ▸ $\setminus$ ⁞ → ⁞ → *Pick points 5 and 6 on the TOP datum* → ↖ → *Select the two straight lines and the arc (Hold "shift" key while selecting multiple items)* → EDIT → MIRROR → *Select the centerline* **Refer Fig. 6.12.**
	25. Modify the dimensions.	↖ → *Double click each dimension and enter the corresponding value* **Refer Fig. 6.12.**
	26. Exit sketcher.	✔
	27. Define the direction and the depth of material removal.	Okay → Blind → Done → 0.1 → ✅
	28. Accept the feature creation after previewing.	Preview → VIEW → DEFAULT ORIENTATION → OK **Refer Fig. 6.13.**

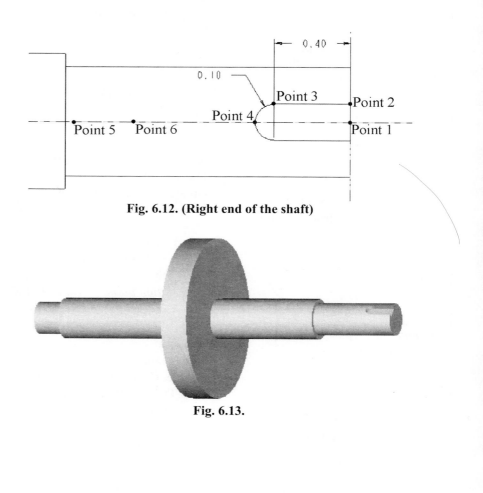

Fig. 6.12. (Right end of the shaft)

Fig. 6.13.

Goal	Step	Commands
Create a hole	29. Start "Hole" feature.	Create → Solid → Hole
	30. Define the parameters.	(Hole Type) Straight hole → (Diameter) 0.375 → (Depth One) *Thru All* → (Primary Reference) *Select the surface shown in Fig. 6.14.* → (Placement Type) *Radial* → (Axial Reference) *Select the shaft axis* → (Distance) 1 → (Angular Reference) *Select the TOP datum plane* (Angle) 45 → ***ENTER*** **Refer Fig. 6.15.**
	31. Create the feature.	✓
Pattern the hole	32. Start "Pattern" command.	Pattern
	33. Select feature to be patterned.	Select → Pick → *Select the hole from the model tree*
	34. Define the pattern parameters.	Identical → Done → Value → *Select the 45° angle* → 90 → ✓ → Done → 4 → ✓ → Done
Resume all features	35. Resume all features.	Resume → All → Done **Refer Fig. 6.16.**
Save the file and exit ProE	36. Save the file and exit ProE.	FILE → SAVE → SHAFT.PRT → ✓ → FILE → EXIT → YES

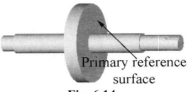

Fig. 6.14.

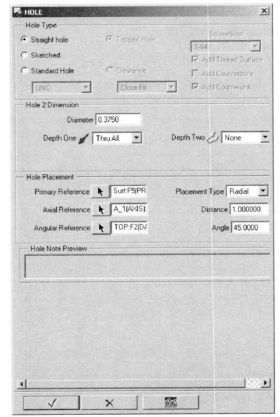

Fig. 6.15.

Fig. 6.16.

LESSON 7
NUTS AND BOLTS

Learning Objectives:

- Use sketches in part creation.

- Create *Cosmetic Threads*.

- Learn the use of **Make Datum** command.

- Practice *Protrusion – Extrude, Cut – Revolve, Cut – Extrude, Chamfer* and *Hole* features.

- Understand the importance of *Relations* and *Dependent Features.*

Design Information:

A bolt is a fastener used for clamping two or more components. Several types of bolt heads are available. A socket head helps in applying large torque using Allen wrenches. If used with a countersunk hole, the bolt head will be flush with the surface.

Nuts incorporate internal threads and engage with bolts. They are normally torqued to a high value to prevent loosening due to vibrations. Special nuts are available with cotter pins, chemical thread locking systems, and lock washers to prevent loosening.

While the regular threads can be created in ProE, it is a common practice to use cosmetic threads instead. They reduce the regeneration time. Also, the key thread parameters such as threads per inch are more easily accessible from the cosmetic thread specification.

Sequence of Steps

Goal I: Create a hexagonal section

1. Create a construction circle.

2. Divide the circle into six segments.

3. Create the hexagon.

Goal II: Create the base feature

1. Append the section to the model.

2. Define the direction and the depth of extrusion.

TOP

FRONT

RIGHT

Goal III: Create a hole

1. Define the diameter of the hole.

2. Select the primary reference surface.

3. Specify the distance from FRONT and RIGHT datum planes.

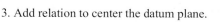

Top surface of the nut

Inside surface of the hole

Bottom surface of the nut (not seen in this view)

Goal IV: Create cosmetic threads

1. Select the thread surface, the start surface and the upto surface

2. Define the thread parameters.

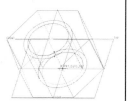

Goal V: Create the cut

1. Sketch the cutting line on the FRONT datum plane.

2. Sketch the axis of revolution and the angle of revolution.

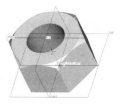

Goal VI: Mirror the cut

1. Create a datum plane in the middle of the nut parallel to the TOP datum plane.

2. Mirror the cut feature about the datum plane.

3. Add relation to center the datum plane.

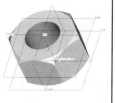

Goal VII: Check the model

1. Check the validity of relation.

2. Check dependent features.

Goal	Step	Commands
Open a new file for the nut section	1. Set up the working directory.	FILE → SET WORKING DIRECTORY → *Select the working directory* → OK
	2. Open a new file.	FILE → NEW → *Sketch* → hexagon → OK
Create the reference coordinate system	3. Create the reference coordinate system.	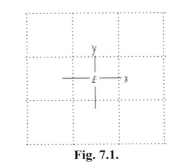 → *Pick a point in the graphics window* **Refer Fig. 7.1.** **The reference coordinate system aids in dimensioning the section.**
Create a hexagon	4. Create a construction circle.	⭕ → *Select the origin of the reference coordinate system* → *Select a point to define the outer edges of the circle* **Refer Fig. 7.2.** 🮲 → *Double click on the diameter dimension* → 1 → ***ENTER*** 🮲 → *Select the circle* → EDIT → TOGGLE CONSTRUCTION **Refer Fig. 7.2.** **"Toggle Construction" changes a regular geometric entity into a construction entity.**

Fig. 7.1.

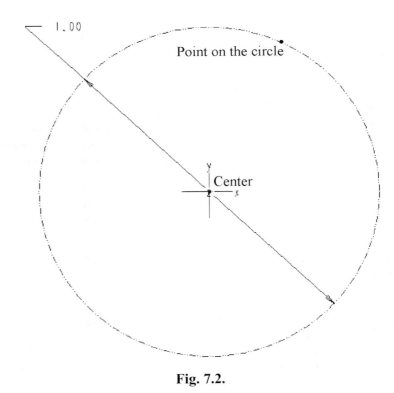

Point on the circle

Center

Fig. 7.2.

Goal	Step	Commands
Create a hexagon (Continued)	5. Create construction lines.	↘ → *Pick points 1 and 2* → *Middle Mouse* → *Pick points 3 and 4 (Horizontal line)* → *Middle Mouse* → *Pick points 5 and 6* → *Middle Mouse* **The lines must pass through the center.** ↖ → *Select lines 1, 2 and 3* (hold SHIFT key while selecting multiple entities) → EDIT → TOGGLE CONSTRUCTION **Refer Fig. 7.3.** ★ **If any entity is shown as a regular geometry, then change it to construction geometry.**
	6. Dimension the inclined lines with respect to the horizontal line.	↦ → *Select the lines between which the angular dimension must be place* → *Middle Mouse to place the angular dimension*
	7. Modify the angular dimensions.	↖ → *Click on the angular dimension* → <u>60</u> → ***ENTER*** → *Click on the second angular dimension* → <u>60</u> → ***ENTER*** **Refer Fig. 7.4.**

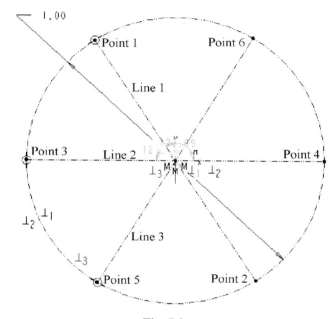

Fig. 7.3.

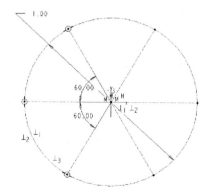

Fig. 7.4.

Goal	Step	Commands
Create a hexagon (Continued)	8. Create the sides of the hexagon.	⬊ → *Click on points 1, 2, 3, 4, 5, 6 and 1* → *Middle Mouse* (to discontinue the line creation) **Refer Fig. 7.5.**
Save the section and exit sketcher	9. Save the section and exit sketcher.	FILE → SAVE → <u>hexagon.sec</u> → ✅ → ✔

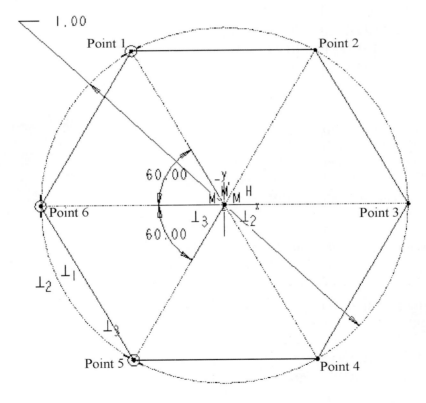

Fig. 7.5.

Goal	Step	Commands
Open a new file for the nut part	10. Open a new file.	FILE → NEW → *Part* → *Solid* → Nut → OK
Create the base feature	11. Start "Protrusion – Extrude" feature.	Feature → Create → Solid → Protrusion → Extrude → Solid → Done → One Side → Done
	12. Set up the sketching plane.	Setup New → Plane → Pick → *Click on the TOP datum plane* → Okay → Right → Plane → Pick → *Click on RIGHT datum plane*
	13. Open the hexagonal section.	SKETCH → DATA FROM FILE → *Select hexagon.sec file* → OPEN
	14. Center the section.	*Drag the section by holding it at the center and dropping it on the PRT_CSYS_DEF (The section coordinate system must lie on the PRT_CSYS_DEF)* → (scale) 0.65 → (rotate) 0 → ✓ → VIEW → REFIT **Refer Figs. 7.6, 7.7 and 7.8.**
	15. Exit sketcher.	✓
	16. Define the depth.	**Blind** → **Done** → 0.35 → ✓ → OK
	17. View the model in the default orientation.	VIEW → DEFAULT ORIENTATION **Refer Fig. 7.9.**

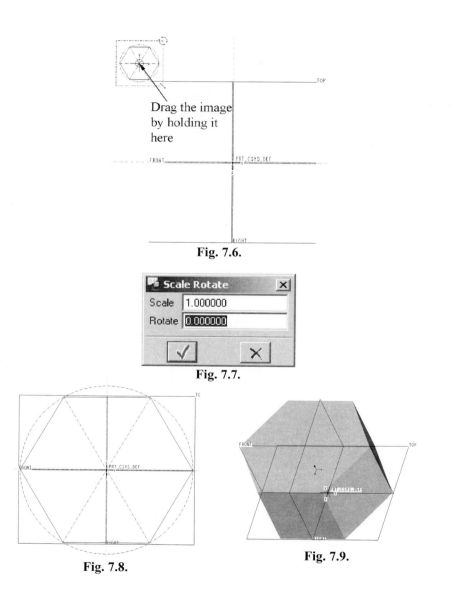

Drag the image by holding it here

Fig. 7.6.

Scale Rotate
Scale 1.000000
Rotate 0.000000

Fig. 7.7.

Fig. 7.8.

Fig. 7.9.

Goal	Step	Commands
Create a hole	18. Create a hole.	Create→ Solid→ Hole **ProE opens the "hole" window.** **Refer Fig. 7.10.** (Hole Type) Straight → (Diameter) 0.313 → (Depth One) *Thru All* → (Primary Reference) ↖ → Query Selec (Right Mouse)→ *Select the top surface of the nut*→ Accept (Middle Mouse) → **Primary reference refers to the plane on which the hole is placed.** (Linear Reference) ↖ → *Select the FRONT datum plane from the model tree* → (Distance) 0 → (Linear Reference) ↖ → *Select the RIGHT datum plane from the model tree* → (Distance) 0 → ☑ **Linear references determine the location of the hole on the placement plane.**

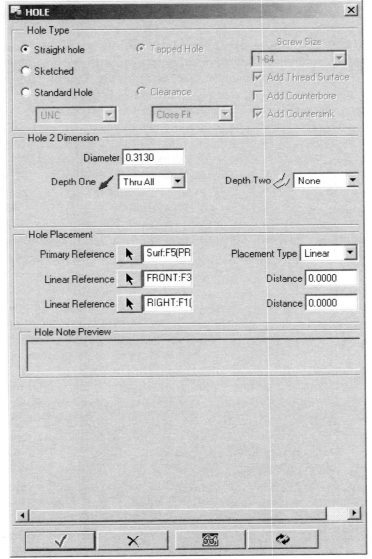

Fig. 7.10.

Goal	Step	Commands
Create a cosmetic thread	19. Start "Cosmetic – Thread" feature.	**In ProE, helical sweep can be used to create threads. However, this feature is both memory intensive and time consuming to regenerate. Therefore, it is a common practice to use a cosmetic feature to create threads. Even though this feature does not show the actual threads, it allows the specification of (and easy access to) various thread parameters.** Create → Cosmetic → Thread → Pick → *Select the inside surface of the hole* → *Select the top surface of the nut* → Okay → Up To Surface → Done → *Query select the bottom surface of the nut* → 0.375 → ✓ **Refer Fig. 7.11.**
	20. Specify the thread parameters.	Mod Params → Enter the parameter values → **Refer Fig. 7.12.** FILE → SAVE → FILE → EXIT → Done/Return
	21. Create the threaded surface.	OK

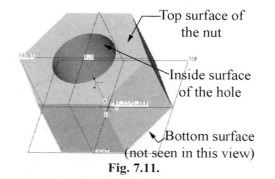

Fig. 7.11.

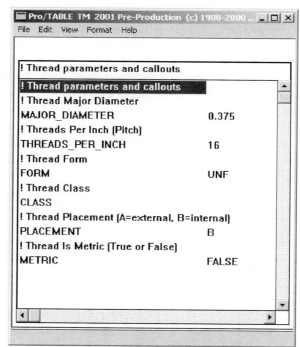

Fig. 7.12.

The term UNF refers to Unified National – Fine thread.

Goal	Step	Commands
Create a cosmetic thread (Continued)	22. View the threaded surface.	VIEW → DEFAULT ORIENTATION → ⊞ **Refer Fig. 7.13.** **Note that the cosmetic features are not visible in the shaded mode.**
Cut top corners	23. Start "Cut – Revolve" feature.	Create → Solid → Cut → Revolve → Solid → Done
	24. Define the direction of revolution.	One Side → Done
	25. Define and orient the sketcher.	Setup New → Plane → Pick → *Select the FRONT datum plane* → Okay → TOP → *Select the TOP datum plane*
	26. Add new references.	**If the references window is not visible, then activate it using SKETCH → REFERENCES command.** *Select the top surface → Select the left edge of the nut* **Refer Figs. 7.14 and 7.15.**

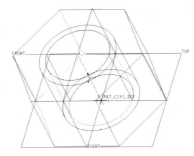

Fig. 7.13.

Fig. 7.14.

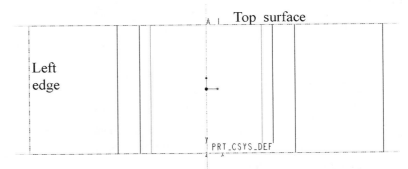

Fig. 7.15.

Goal	Step	Commands
Cut top corners (Continued)	27. Sketch the cutting line.	⬉ → *Pick points 1 and 2* **Refer Fig. 7.16.**
	28. Dimension the line.	↔ → *Select the line* → *Select the top surface* → *Middle Mouse to place the angular dimension* ↖ → *Double click on the distance dimension* → 0.045 → **ENTER** → *Double click on the angular dimension* → 45 → **ENTER** **Refer Fig. 7.16.**
	29. Sketch the axis of revolution on the RIGHT datum plane.	⬉ ⬉ ⋮ → ⋮ → *Pick points 3 and 4 on the RIGHT datum plane* **Refer Fig. 7.16.**
	30. Exit sketcher.	✔
	31. Define the direction of material removal.	Flip → Okay **Note that the arrow should point outwards.**
	32. Define the angle of revolution.	360 → Done
	33. Accept the feature creation after previewing.	Preview → VIEW → DEFAULT ORIENTATION → OK **Refer Fig. 7.17.**

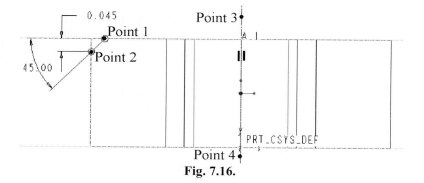

Fig. 7.16.

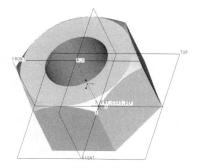

Fig. 7.17.

Goal	Step	Commands
Mirror the cut geometry	34. Select the feature to be mirrored.	Copy → Mirror → Select → Dependent → Done → *Select the cut feature from the model tree* → Done
	35. Create a datum plane for mirroring.	**We can create datum planes on the fly using the "Make Datum" command. This datum plane is used for mirroring operation.** Make Datum → Offset → Plane → Coor Sys → Pick → *Select the TOP datum plane* → Enter Value → 0.175 → ✓ → Done **Refer Fig. 7.18.**
	36. Add relation.	▶ PART → Relations → Part Rel → Pick → *Click on DTM 1* → *Click on base feature (First protrusion) from the model tree* **Refer Fig. 7.19.** **ProE displays the names of variables. The names may vary depending on the order of feature creation. In our case, d10 must be half of d3. These variables may have slightly different names in your model.** Add → d10 = 0.5 * d3 → ✓ → ✓ → Done

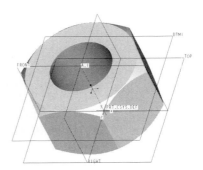

Fig. 7.18.

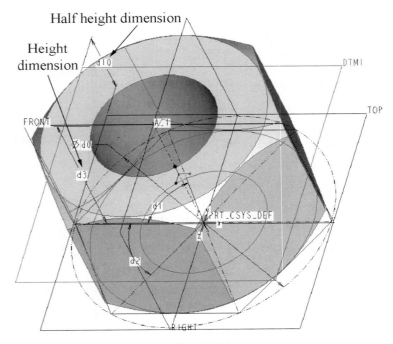

Half height dimension

Height dimension

Fig. 7.19.

Goal	Step	Commands
Check the model	37. Check the validity of the relationship and dependent features.	Modify → *Double click on the height dimension of the nut* → 1 → ✓ → Regenerate **By adding the relationship (d12 = 0.5 * d5), the model is flexible, i.e., it is possible to quickly change the height of the nut. The mirror feature would regenerate without problems.** Modify → *Double click on the height dimension of the cut* → 0.1 → ✓ → Regenerate **By making the bottom cut dependent on the top cut, the model is flexible, i.e., it is possible to change the height of the both cuts by changing one of them.** **Refer Fig. 7.20.** Modify → *Double click on the height dimension of the nut* → 0.375 → ✓ → Regenerate → Modify → *Double click on the height dimension of the cut* → 0.045 → ✓ → Regenerate
Save the file and close the window	38. Save the file and close window.	FILE → SAVE → NUT.PRT → ✓ → FILE → ERASE → CURRENT

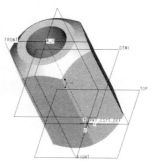

Fig. 7.20.

Sequence of Steps

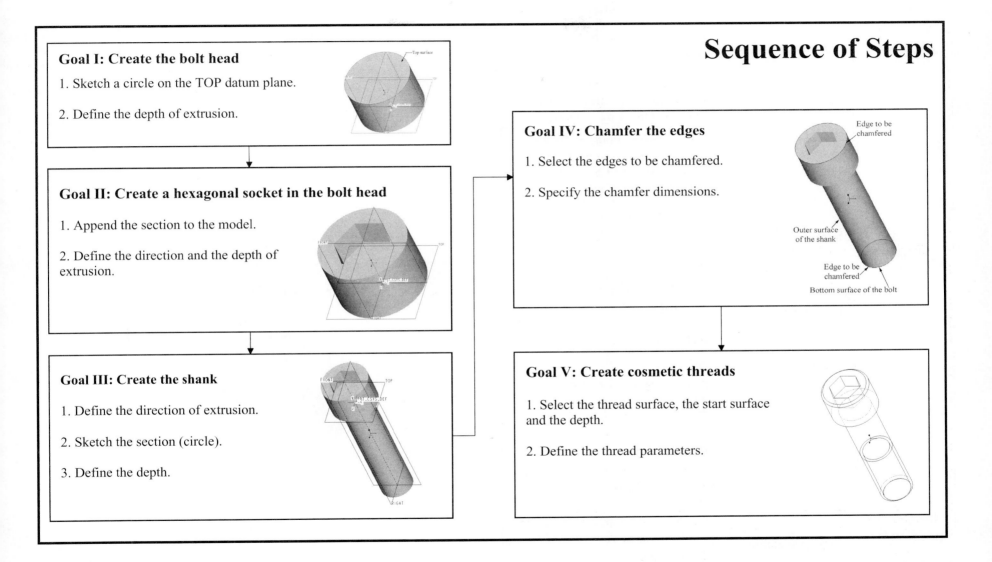

Goal I: Create the bolt head

1. Sketch a circle on the TOP datum plane.

2. Define the depth of extrusion.

Goal II: Create a hexagonal socket in the bolt head

1. Append the section to the model.

2. Define the direction and the depth of extrusion.

Goal III: Create the shank

1. Define the direction of extrusion.

2. Sketch the section (circle).

3. Define the depth.

Goal IV: Chamfer the edges

1. Select the edges to be chamfered.

2. Specify the chamfer dimensions.

Goal V: Create cosmetic threads

1. Select the thread surface, the start surface and the depth.

2. Define the thread parameters.

Goal	Step	Commands
Open a new file for the bolt part	1. Open a new file.	FILE → NEW → *Part* → *Solid* → Bolt → OK
Create the bolt head	2. Start "Protrusion – Extrude" feature.	Feature → Create → Solid → Protrusion → Extrude → Solid → Done → One Side → Done
	3. Set up the sketching plane.	Setup New → Plane → Pick → *Select the TOP datum plane* → Okay → Right → *Select the RIGHT datum plane*
	4. Create a circular section.	O → *Select the center and then, a point to define the outer edge of the circle*
	5. Modify the dimensions.	↖ → *Double click on the diameter dimension* → 0.55 → ***ENTER*** **Refer Fig. 7.21.**
	6. Exit sketcher.	✔
	7. Define the depth.	Blind → Done → 0.375 → ✔ → OK → VIEW → DEFAULT ORIENTATION **Refer Fig. 7.22.**
Create a hexagonal socket in the head	8. Start "Cut – Extrude " feature.	Create → Solid → Cut → Extrude → Solid → Done → One Side → Done
	9. Set up sketching plane.	Setup New → Plane → Pick → *Click on top surface of the bolt head* → Okay → Right → Plane → Pick → *Click on RIGHT datum*

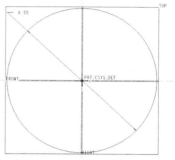

Fig. 7.21.

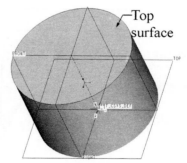

Top surface

Fig. 7.22.

Goal	Step	Commands
Create a hexagonal socket in the head (Continued)	10. Insert the hexagonal section.	SKETCH → DATA FROM FILE → *Select hexgon.sec file* → OPEN
	11. Center the section.	*Drag the section by holding it at the center and dropping it on the PRT_CSYS_DEF (The section coordinate system must lie on the PRT_CSYS_DEF)* → (scale) 0.5 → (rotate) 0 → ✓ → VIEW → REFIT **Refer Fig. 7.23.**
	12. Delete the diameter dimension.	↖ → *Select the diameter dimension* → **DELETE**
	13. Add a new dimension (distance between two parallel sides).	↦ → *Select the top horizontal line and then, the bottom horizontal line of the hexagonal bolt* → *Middle Mouse to place the dimension* **Refer Fig. 7.24.**
	14. Modify the width.	↖ → *Double click on the width dimension* → 5/16 → **ENTER**
	15. Exit sketcher.	✔

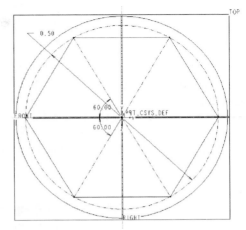

Fig. 7.23.

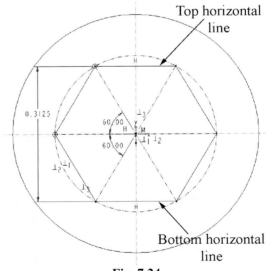

Top horizontal line

Bottom horizontal line

Fig. 7.24

Goal	Step	Commands
Create a hexagonal socket in the head (Continued)	16. Define the depth.	Okay → Blind → Done → 0.182 → ✓ → OK
	17. View the model in the default view.	VIEW → DEFAULT ORIENTATION **Refer Fig. 7.25.**
Create the shank	18. Start "Protrusion – Extrude" feature.	Create → Solid → Protrusion → Extrude → Solid → Done → One Side → Done
	19. Set up the sketching plane.	Setup New → Plane → Pick → *Select the TOP datum plane* → Flip → Okay → Right → *Select the RIGHT datum plane*
	20. Create a circular section.	O → *Select the center and then, a point to define the circle*
	21. Modify the dimensions.	↖ → *Double click on the diameter dimension* → 0.375 → *ENTER* **Refer Fig. 7.26.**
	22. Exit sketcher.	✔
	23. Define the depth.	Blind →Done→ 1.5 → ✓ → OK
	24. View the model in the default orientation.	VIEW → DEFAULT ORIENTATION **Refer Fig. 7.27.**

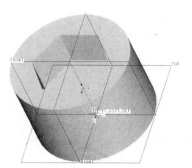

Fig. 7.25.

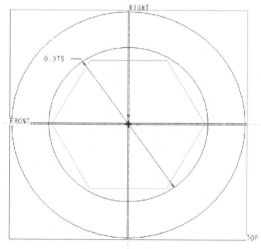

Fig. 7.26.

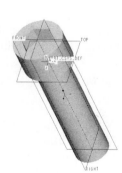

Fig. 7.27.

Goal	Step	Commands
Chamfer the edges	25. Chamfer the top and bottom edges of the bolt.	Create → Solid → Chamfer → Edge → 45 X D → 0.025 → ✓ → Pick → *Select the top edge of the head and the free end of the shank* → Done Sel → Done Ref → OK **Refer Figs. 7.28.**
Create a cosmetic thread	26. Create cosmetic threads.	Create → Cosmetic → Thread → Pick → *Select the outer surface of the shank* → *Select the bottom surface of the bolt (End of the bolt)* → Okay → Blind → Done → 0.75 → ✓ → 0.325 → ✓ **Refer Fig. 7.28.**
	27. Specify thread parameters.	Mod Params → Enter the parameter values → **Refer Fig. 7.29.** FILE → SAVE → FILE → EXIT → Done/Return
	28. Create the threaded surface.	OK
	29. View the threaded surface.	VIEW → DEFAULT ORIENTATION → ⊞ **Refer Fig. 7.30.**
Save the file and exit ProE	30. Save the file and exit ProE.	FILE → SAVE → BOLT.PRT → ✓ → FILE → EXIT → YES

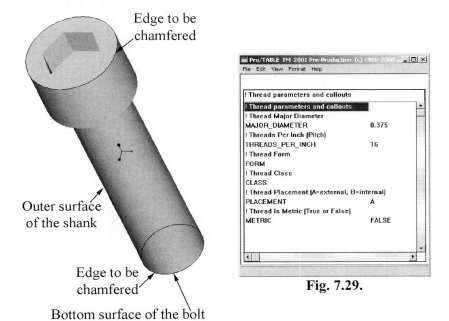

Edge to be chamfered

Outer surface of the shank

Edge to be chamfered

Bottom surface of the bolt

Fig. 7.28.

Fig. 7.29.

Fig. 7.30.

LESSON 8
BOLT HEADS

Learning Objectives:

- Understand *Family Tables* and their applications in design.

- Use *Sketches* in part creation.

- Practice *Relations*.

- Practice *Protrusion – Extrude*.

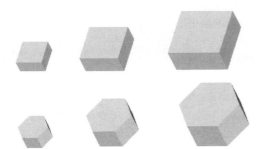

Design Information:

Family tables provide a systematic means for organizing and storing a large database of information about similar components. The user must capture the common geometric characteristics of a set of similar components in order to create a generic instance. Then, the user can create specific instances by defining the dimensions of the geometric features.

A good application for family tables is the selection of bolts. While the head may be different, the shank and the thread specification are common characteristics of any bolt. The dimensions of these features may vary. Family tables allow the user to create a single database of bolts. The user can then create a specific instance by selecting the type of bolt head and specifying the relevant dimensions.

Sequence of Steps

Goal I: Create a hexagonal section

1. Create a construction circle.

2. Divide the circle into six segments.

3. Create the hexagon.

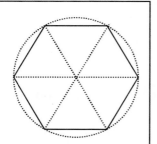

Goal II: Create a square section

1. Create a square section.

2. Add relations to center the square.

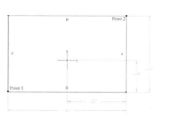

Goal III: Create the hexagonal head

1. Import the hexagonal section.

2. Define the direction and the depth of extrusion.

3. Name the hexagonal head feature and its dimensions.

Goal IV: Create the square head

1. Import the square section.

2. Define the direction and the depth of extrusion.

3. Name the square head feature and its dimensions.

Goal V: Create the family table

1. Add features and dimensions to the family table.

2. Add new instances to the family table.

3. Patternize the family table to create several instances.

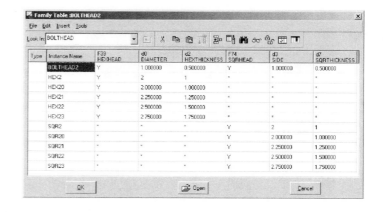

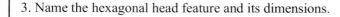

Goal	Step	Commands
Open a new file for the hexagonal head section	1. Set up the working directory.	FILE → SET WORKING DIRECTORY → *Select the working directory* → OK
	2. Open a new file.	FILE → NEW → *Sketch* → hexhead → OK
Create the reference coordinate system	3. Create the reference coordinate system.	 → *Click in the graphics window* **Refer Fig. 8.1.** **The reference coordinate system aids in dimensioning the section.**
Create a hexagon	4. Create a construction circle.	⭕ → *Pick the center at the origin of the reference coordinate system* → *Select a point to define the outer edge of the circle* → ↖ → *Double click the diameter dimension* → 1 → **ENTER** → ↖ → *Select the circle* → EDIT → TOGGLE CONSTRUCTION **Refer Fig. 8.2.** **"Toggle construction" changes a regular geometric entity into a construction entity.**

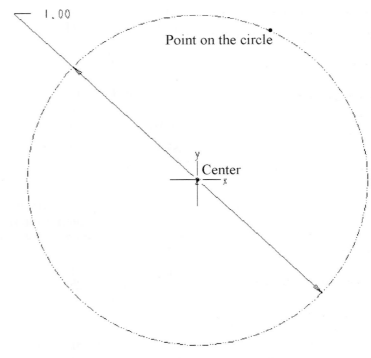

Fig. 8.1.

Fig. 8.2.

Goal	Step	Commands
Create a hexagon (Continued)	5. Create the construction lines.	＼ → *Pick points 1 and 2* → *Middle Mouse* → *Pick points 3 and 4* → *Middle Mouse* → *Pick points 5 and 6* → *Middle Mouse* **The lines must pass through the center.** ➤ → *Select lines 1, 2 and 3* (hold SHIFT key while selecting multiple entities) → EDIT → TOGGLE CONSTRUCTION **Refer Fig. 8.3.** **If any entity is shown in regular geometry, then change it to construction geometry.**
	6. Align points 1 and 6.	SKECTH → CONSTRAIN or 〔⫱〕 **ProE opens the "constraints" window. For explanation of constraints, refer Fig. 8.4.** ↔ → *Select points 1 and 6*
	7. Dimension the inclined lines with respect to the horizontal line.	⟷ → *Select line 1* → *Select line 2* → *Middle Mouse* (between the lines to place the angular dimension)

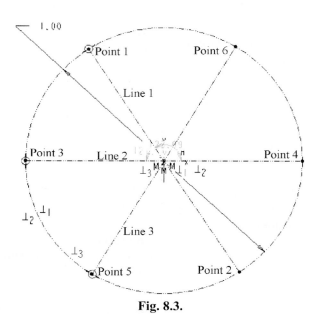

Fig. 8.3.

↕ Makes line or two vertices vertical	↔ Makes line or two vertices horizontal	⊥ Makes two entities perpendicular
⊘ Makes two entities tangential	＼ Places point on the middle of the line	⊙ Creates same points, points on entity or collinear constraint
⇥⇤ Makes two points or vertices symmetric about the centerline	= Makes the lengths, radii or curvatures equal	// Makes two lines parallel

Fig. 8.4.

Goal	Step	Commands
Create a hexagon (Continued)	8. Modify the angular dimensions.	↖ → *Double click the angular dimension* → 60 → ✓ **Refer Fig. 8.5.**
	9. Create the sides of the hexagon.	＼ → *Pick points 1, 2, 3, 4, 5, 6 and 1* → *Middle Mouse* (to discontinue the line creation) **Refer Fig. 8.5.**
Save the sketch and exit sketcher	10. Save the section and exit sketcher.	FILE → SAVE → hexhead.sec → ✓ → ✓

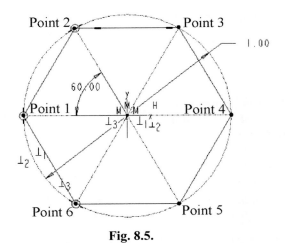

Fig. 8.5.

Goal	Step	Commands
Open a new file	11. Open a file for the square section.	FILE → NEW → *Sketch* → sqrhead → OK
Create the reference coordinate system	12. Create the reference coordinate system.	→ *Click in the graphics window* **Refer Fig. 8.6.**
Create a square section	13. Create a rectangle.	□ → *Pick points 1 and 2* **Refer Fig. 8.7.**
	14. Add relations to center the geometry.	SKETCH → RELATIONS **The following equations may be slightly different for you.** Add → sd5 = 0.5* sd3 → ✓ → sd4 = 0.5 * sd2 → ✓ → sd3=sd2 → ✓ → ✓
	15. Sort relations.	Sort Rel **The sort relations command orders the relations so that a relation that depends on the value of another relation is evaluated after that relation. It also detects circular relations.**
	16. Modify the dimensions.	↖ → *Double click the sd2 dimension* → 1→ **ENTER**
Save the section and exit sketcher	17. Save the section and exit sketcher.	FILE → SAVE → sqrhead.sec → ✓ → ✔

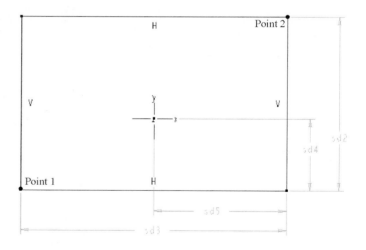

Fig. 8.6.

Fig. 8.7.

Goal	Step	Commands
Open a new file for the bolt head part	18. Open a new file.	FILE → NEW → *Part* → Solid → bolthead → OK
Create the hex head	19. Start "Protrusion – Extrude" feature.	Feature → Create → Solid → Protrusion → Extrude → Solid → Done → One Side → Done
	20. Set up sketching plane.	Setup New → Plane → Pick → *Select the TOP datum plane* → Okay → Right → Plane → Pick → *Select the RIGHT datum plane*
	21. Insert the hexagonal section.	SKETCH → DATA FROM FILE → *Select hexhead.sec file* → OPEN
	22. Center the section.	Drag the section by holding it at the center and drop it on the PRT_CSYS_DEF (The section coordinate system must lie on the PRT_CSYS_DEF) **Refer Fig. 8.8.**
	23. Define the scale.	(scale) 1 → (rotate) 0 → ✓ → VIEW → REFIT **Refer Fig. 8.9.**
	24. **Optional Step:** If the diameter of the circle is not 1, then change it to 1	▶ → *Double click the diameter dimension* → 1 → ***ENTER***

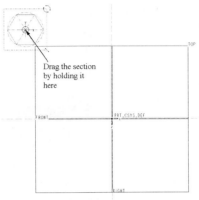

Fig. 8.8.

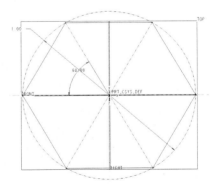

Fig. 8.9.

Goal	Step	Commands
Create the hex head (Continued)	25. Exit sketcher.	**At this point, the section should be centered around the PRT_CSYS_DEF.** ✔
	26. Define the depth.	**Blind** → **Done** → <u>0.5</u> → ✔ → OK
	27. View the model in the default orientation.	VIEW → DEFAULT ORIENTATION **Refer Fig. 8.10.**
	28. Rename the feature.	▶ PART (to expand part menu)→ Set Up → Name → Feature → Pick → *Select the protrusion feature from the model tree* → <u>HexHead</u> → ✔ → Done Sel → Done
	29. Rename the dimensions.	Modify → Dimension → Pick → *Select the HexHead from the model tree* → *Select the diameter (1.0) dimension* → Done Sel → **ProE opens Modify Dimension window.** *Select the Dimension Text tab* → (Name) <u>Diameter</u> → OK → **Refer Fig. 8.11.** *Select the thickness (0.5) dimension* → Done Sel → *Select the Dimension Text tab* → (Name) <u>HexThickness</u> → OK

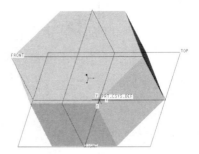

Fig. 8.10.

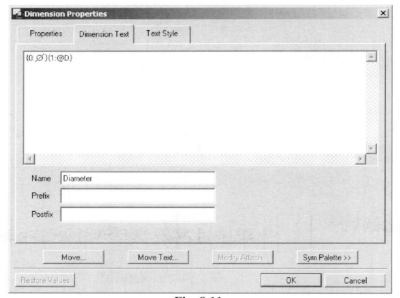

Fig. 8.11.

Goal	Step	Commands
Suppress the hex head	30. Suppress the hex head feature.	Feature → Suppress → Normal → Select → Pick → *Select the HexHead feature from the model tree* → Done
Create a square head	31. Start "Protrusion – Extrude" feature	Create → Solid → Protrusion → Extrude → Solid → Done → One Side → Done
	32. Set up sketching plane.	Setup New → Plane → Pick → *Select the TOP datum plane* → Okay → Right → Plane → Pick → *Select the RIGHT datum plane*
	33. Insert the square section.	SKETCH → DATA FROM FILE → *Select sqrhead.sec file* → OPEN
	34. Center the section.	Drag the section by holding it at the center and drop it on the PRT_CSYS_DEF (The section coordinate system must lie on the PRT_CSYS_DEF) **Refer Fig. 8.12.**
	35. Define the scale.	(scale) 1 → (rotate) 0 → ✓ → VIEW → REFIT
	36. Optional Step: **Make sure that the side is 1.** If not, change the dimension.	↖ → *Double click the side dimension* → 1 → ✓
	37. Exit sketcher.	✓
	38. Define the depth.	**Blind** → **Done** → 0.5 → ✓ → OK
	39. View the model in the default orientation.	VIEW → DEFAULT ORIENTATION **Refer Fig. 8.13.**

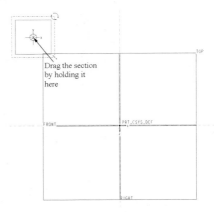

Fig. 8.12.

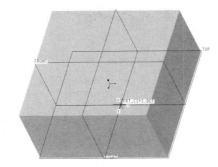

Fig. 8.13.

Goal	Step	Commands
Create a square head (Continued)	40. Rename the feature.	▶ PART (to expand part menu)→ Set Up → Name → Feature → Pick → *Select the protrusion feature from the model tree* → SqrHead → ✓ → Done
	41. Rename the side dimension.	Modify → Value → Pick → *Select the SqrHead from the model tree → Try to select 1.0 dimension. As a relationship governs the two 1.0 dimensions, ProE allows you to select only one of them→* 1 → ✓ → Dimension → Pick → *Select the 1.0 dimension (previously modified) →* Done Sel → *Select the Dimension Text tab →* (Name) Side → OK **Refer Fig. 8.14.** *Select the thickness (0.50) dimension →* Done Sel → *Select the Dimension Text tab →* (Name) SqrThickness → OK
Resume all features	42. Resume the hex head feature.	Feature → Resume → All → Done → Done **HexHead feature should reappear in the model tree.** **It is not visible as it is contained within the SqrHead feature.**

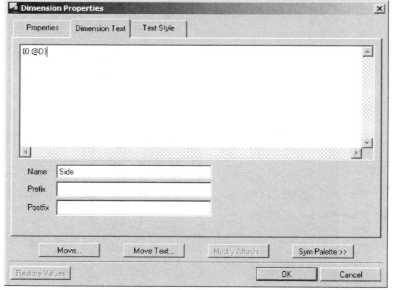

Fig. 8.14.

Goal	Step	Commands
Create the family table	43. Start the Family Table command.	Family Tab **ProE opens the Family Table window.** **Refer Fig. 8.15.**
	44. Add features and dimensions to the family table.	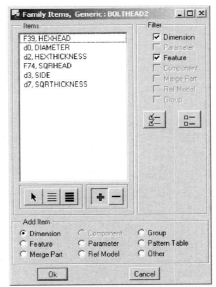 **ProE opens the Family Items window.** *(In the Add Item subwindow)* *Select feature →* Select *→* Pick *→ Select HexHead feature from the Model Tree →* *Select dimension →* Pick *→* *Select HexHead feature → Select the diameter and thickness dimensions →* *Select feature →* Select *→* Pick *→ Select SqrHead feature from the Model Tree →* *Select dimension →* Pick *→* *Select SqrHead feature → Select the side dimension → Select the thickness dimension → Done Sel* *→* OK **Refer Fig. 8.16.**

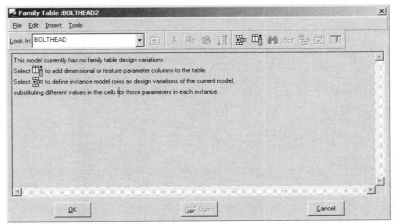

Fig. 8.15.

Fig. 8.16.

Goal	Step	Commands
	45. Add new instances to the table.	![icon] → (Instance Name) Hex2 → (HexHead) Y → (Diameter) 2 → (HexThickness) 1 ![icon] → (Instance Name) Sqr2 → (SqrHead) Y → (Side) 2 → (SqrThickness) 1 **Refer Fig. 8.17.**
Create the family table (Continued)	46. Create more instances (patternize the table).	*Select Hex2* → ![icon] → (Quantity) <u>4</u> → *Select diameter* → *Select HexThickness* → <u>>></u> → *Select the diameter* →(Increment) <u>0.25</u> → ***ENTER*** → *Select the HexThickness* → (Increment) <u>0.25</u> → ***ENTER*** → OK → **Refer Fig. 8.18.** *Select SQRB2* → ![icon] → (Quantity) <u>4</u> → <u>Select Side parameter</u> → <u>Select SqrThickness</u> → <u>>></u> → *Select the side* → (Increment) <u>0.25</u> → ***ENTER*** → *Select the SqrThickness* → (Increment) <u>0.25</u> →***ENTER*** → OK → **Refer Fig. 8.19.** OK **Refer Fig. 8.20.**

Fig. 8.17.

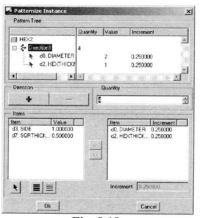

Fig. 8.18.

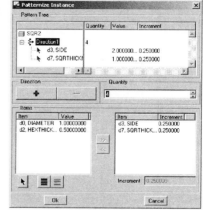

Fig. 8.19.

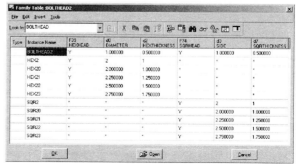

Fig. 8.20.

Goal	Step	Commands
Save the file and exit the window	47. Save the file and erase the window.	FILE → SAVE → BOLTHEAD.PRT → ✓ → FILE → ERASE → CURRENT → YES
Open the bolt file	48. Open the bolt file.	FILE → OPEN → *Select BOLTHEAD.PRT → Select the instance* → OPEN **Note that it is possible to select an instance by choosing the values for different parameter specification by clicking on the Parameter Tab.**
Exit ProE	49. Exit ProE.	FILE → EXIT → YES

Learning Objectives:

- Create **Protrusion – Blend**.

- Practice **Relations** and **Mirror** commands.

- Practice importing sketches.

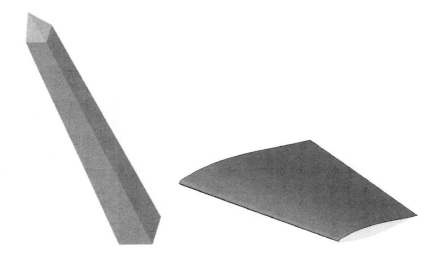

Design Information:

Washington Monument was designed by Robert Mills to pay tribute to George Washington's achievements. It follows the form of an obelisk. With a height of 550 ft. and a weight of 90,000 tons, it is the largest masonry structure in the world. In 30-mile wind gusts, the sway of the structure is less than 0.125 inches. In this lesson, the monument will be modeled by using the blend feature.

Wings are designed to maximize the lift created due to the airflow. Several factors such as the air density, speed and wing area affect the forces on the wing: lift and drag. Based on these factors, the wing sections are determined at regular intervals along the length of the wing. The sections are often defined by a set of data points. These sections are then joined using straight blend feature.

Goal I: Set up units

1. Change the units to FPS system.

Sequence of Steps

Goal II: Create protrusion blend feature

1. Define the blend parameters as parallel and regular sections.

2. Select the type of transition surface as straight.

3. Define and orient the sketching plane.

4. Sketch the ground section (section #1) of the obelisk.

5. Toggle section.

6. Sketch the second section (section #2) of the obelisk.

7. Toggle section.

8. Define the tip of the obelisk (section #3)

9. Define the distance between the sections.

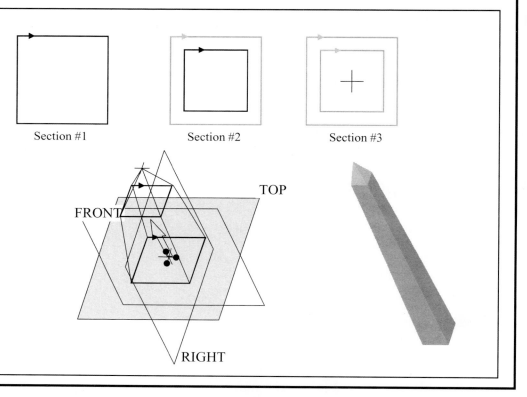

Section #1   Section #2   Section #3

Goal	Step	Commands
Open a new file for the monument part	1. Set up the working directory.	FILE → SET WORKING DIRECTORY → *Select the working directory* → OK
	2. Open a new file.	FILE → NEW → *Part* → *Solid* → monument → OK

Goal	Step	Commands
Set up units	3. Set the units to the FPS system.	Set Up → Units → *Select Foot Pound Second (FPS)* → → Set... → **Refer Fig. 9.1.** OK → CLOSE → Done **Refer Fig. 9.2.**
Create the obelisk	4. Start "Protrusion – Blend" feature.	Feature → Create → Solid → Protrusion → Blend → Solid → Done
	5. Select the sketch option.	Parallel → Regular Sec → Sketch Sec → Done
	6. Select the type of transition surface.	Straight → Done **Straight option connects the vertices of various sections using straight lines whereas smooth connects the vertices with curves.**
	7. Select the sketching plane.	Setup New → Plane → Pick → *Select the TOP datum plane*
	8. Define the direction of material extrusion.	Okay
	9. Orient the sketcher.	Right → Plane → Pick → *Select the RIGHT datum plane*

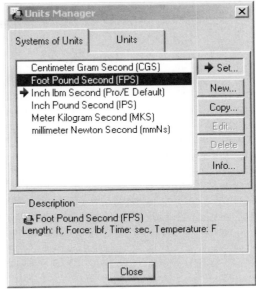

Fig. 9.1.

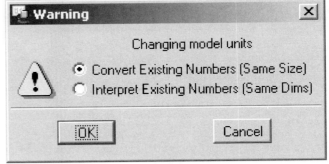

Fig.9.2.

In parallel blend, the sections that are blended are parallel and the depth is specified. On the other hand, the sections are separated by specified angle in the rotational blend option.

Goal	Step	Commands
Create the obelisk (Continued)	10. Sketch a rectangle.	□ → *Pick points 1 and 2* **Refer Fig. 9.3.**
	11. Add relations to center the section.	**Note that sd#s may be slightly different in your model.** SKETCH → RELATIONS → Add → sd5 = 0.5* sd3 → ✓ → sd4 = 0.5 * sd2 → ✓ → sd3 = sd2 → ✓ → ✓
	12. Sort relations.	Sort Rel **The "Sort Rel" command sorts the relations in the order of precedence.**
	13. Modify the side dimension.	↖ → *Double click the side sd2 dimension* → 55 → ***ENTER*** **Refer Fig. 9.4.**
	14. Toggle the section to construct a new section.	SKETCH → FEATURE TOOLS → TOGGLE SECTION

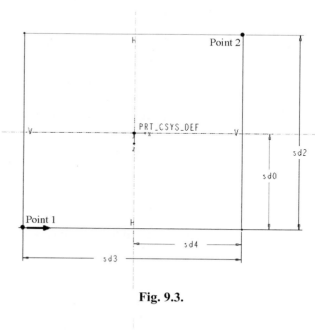

Fig. 9.3.

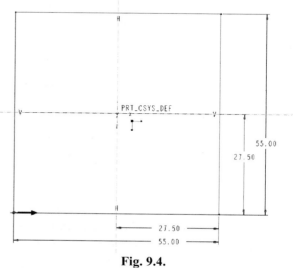

Fig. 9.4.

Goal	Step	Commands
Create the obelisk (Continued)	15. Sketch a rectangle.	▢ → *Pick points 3 and 4* **Refer Fig. 9.5.**
	16. Add dimensions.	⟷ → *Select the bottom side → Middle Mouse to place the horizontal dimension →* *Select the right side → Middle Mouse to place the vertical dimension →* *Select the top side → Select the FRONT datum plane → Middle Mouse to place the vertical dimension →* *Select the right vertical side → Select the RIGHT datum plane → Middle Mouse to place the horizontal dimension* **Refer Fig. 9.5.**
	17. Add relations to center the section.	**Note that sd#s may be slightly different in your model.** SKETCH → RELATIONS → Add → sd14 = 0.5* sd11 → ✓ → sd13 = 0.5 * sd12 → ✓ → sd12 = sd11 → ✓ → ✓
	18. Sort relations.	Sort Rel **The "Sort Rel" command sorts the relations in the order of precedence.**

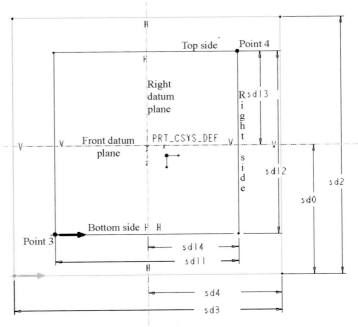

Fig. 9.5. The first section is shown in gray color.

Goal	Step	Commands
Create the obelisk (Continued)	19. Modify the side dimension.	↖ → *Double click the sd11 dimension* → 34.5 → ***ENTER*** **Refer Fig. 9.6.**
	20. Toggle the section to construct new section.	SKETCH → FEATURE TOOLS → TOGGLE SECTION
	21. Sketch a data point.	[icons] → [icon] → *Pick point 5 at the intersection of the FRONT and RIGHT datum planes* **Refer Fig. 9.6.**
	22. Exit sketcher.	✔
	23. Define the depths.	500 → ✔ → 55 → ✔
	24. Accept the feature creation.	OK → VIEW → DEFAULT ORIENTATION **Refer Fig. 9.7.**
Save the file and exit ProE	25. Save the file and exit ProE.	FILE → SAVE → MONUMENT.PRT → ✔ → FILE → EXIT

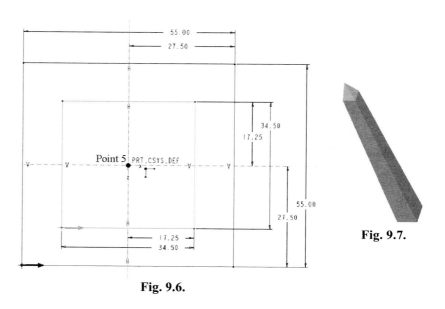

Fig. 9.7.

Fig. 9.6.

ProE connects the corresponding vertices of the two sections starting with the start points. The start point can be varied by "SKETCH → FEATURE TOOLS → START POINT" command. We can move from one section to another by using "SKETCH → FEATURE TOOLS → TOGGLE SECTION" command.

Sequence of Steps

Goal I: Create the data points

1. Create the data points in Notepad or Excel.

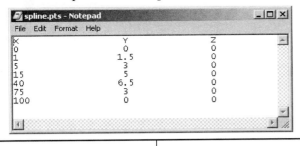

Goal II: Create the airfoil section

1. Create a coordinate system.

2. Create a spline.

3. Assign the coordinate system to the spline.

4. Read data points.

5. Mirror the upper spline to create the lower spline.

Goal III: Create the airfoil

1. Define the blend parameters as parallel and regular sections.

2. Select the type of transition surface as straight.

3. Define and orient the sketching plane.

4. Insert the base section of the wing.

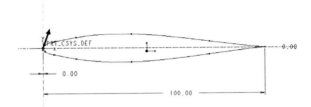

5. Toggle section.

6. Sketch the tip section of the wing.

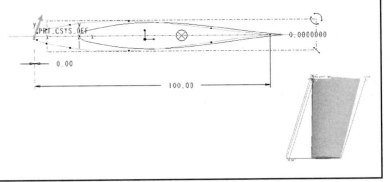

Goal	Step	Commands
IN NOTEPAD OR EXCEL		
Create the airfoil data points	1. Open notepad (or VI editor) or Excel.	
	2. Enter the surface data points.	<u>Enter data points</u> **Refer Fig. 9.8.**
	3. Save the data file.	FILE → SAVE AS → Select the working directory → (File name) <u>"spline.pts"</u> **(If the name is enclosed in quotes, then Windows will not attach any extension to the file name.)** → (Save as type) ANSI → Save FILE → EXIT **Refer Fig. 9.9.**
IN PROE		
Open a new file for the airfoil section	4. Setup the working directory.	FILE → SET WORKING DIRECTORY → *Select the working directory* → OK
	5. Open a new file.	FILE → NEW → Sketch → <u>airfoil</u> → OK
Create the airfoil section	6. Create the coordinate system.	→ *Select a point in the graphics window*
	7. Create a spline.	→ *Pick points 1, 2 and 3* → *Middle Mouse* **Refer Fig. 9.10.**
	8. Modify the dimensions.	→ *Double click each dimension and enter the corresponding value* **Refer Fig. 9.10.**

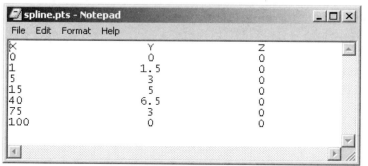

Fig. 9.8.

Fig. 9.9.

Fig. 9.10.

Goal	Step	Commands
Create the airfoil section (Continued)	9. Select the Cartesian coordinate system.	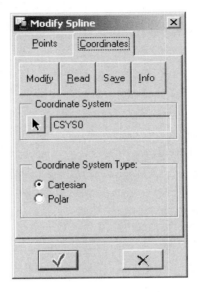 → *Select the spline → Select the Coordinates tab in "Modify Spline" window → Select the Cartesian coordinate system* **Refer Fig. 9.11.**
	10. Assign the spline to the local coordinate system.	*Select the ▶ in the "Modify Spline" window → Query select the coordinate system created in step 9* **The message window should display: Spline is dimensioned to the local coordinate system.** **Refer Fig. 9.11.**
	11. Read data points.	Read → **If ProE prompts: Cannot modify spline with dimensions to internal points. Delete dimensions? Click on** YES *Select "SPLINE.PTS" →* OPEN *→* YES *→* ✓ **Refer Fig. 9.12.**
	12. Create a centerline to mirror the spline.	↘ ↘ ┊ → ┊ → *Pick points 1 and 3* **Refer Fig. 9.12.**

Fig. 9.11.

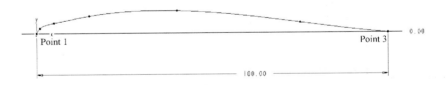

Fig. 9.12.

Goal	Step	Commands
Create the airfoil (Continued)	13. Mirror the spline.	↖ → *Select the spline* → EDIT → MIRROR → *Select the centerline* **Refer Fig. 9.13.**
	14. Save the section file.	FILE → SAVE → AIRFOIL.SEC → ✓ → FILE → CLOSE WINDOW → Yes
Open a new file for the wing part	15. Open a new file.	FILE → NEW → Part → Solid → wing → OK
Create the wing part	16. Start "Protrusion – Blend" feature.	Feature → Create → Solid → Protrusion → Blend → Solid → Done
	17. Select the sketch option.	Parallel → Regular Sec → Sketch Sec → Done
	18. Select the type of transition surface.	Straight → Done
	19. Select the sketching plane.	Setup New → Plane → Pick → *Select the FRONT datum plane*
	20. Define the direction of material extrusion.	Okay
	21. Orient the sketcher.	Right → Plane → Pick → *Select the RIGHT datum plane*
	22. Import the airfoil section.	SKETCH → DATA FROM FILE → *Select "airfoil.sec" file* → *Drag and drop the section so that the section coordinate system is on the PRT_CSYS_DEF* → (Scale) 1 → (Angle) 0 → ✓ **Refer Fig. 9.14.**

Fig. 9.13.

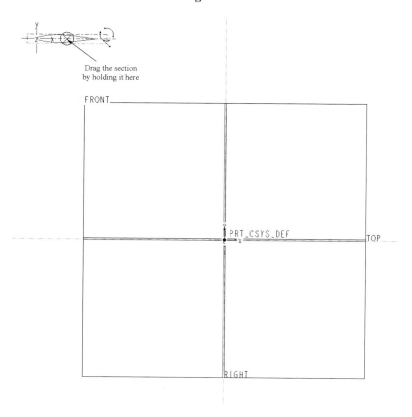

Fig. 9.14.

Goal	Step	Commands
Create the wing part (Continued)	23. Modify the distance between the coordinate systems.	*Double click on the distance dimension* → 0 → ✓ **Refer Fig. 9.15.**
	24. Toggle the section.	SKETCH → FEATURE TOOLS → TOGGLE SECTION **The current section should change to gray color.**
	25. Import the airfoil section.	SKETCH → DATA FROM FILE → *Select "airfoil.sec" file* → *Drag the section and drop it* → **Refer Fig. 9.16.** (Scale) 0.6 → (Angle) 0 → ✓
	26. Modify the placement dimensions.	↖ → *Modify the vertical and horizontal dimensions of the section from the coordinate system* **Refer Fig. 9.17.**
	27. Exit sketcher.	✓
	28. Define the depths.	200 → ✓
	29. Accept the feature creation.	OK → VIEW → DEFAULT ORIENTATION **Refer Fig. 9.18.**
Save the file and exit ProE	30. Save the file and exit ProE.	FILE → SAVE → WING.PRT → ✓ → FILE → EXIT → YES

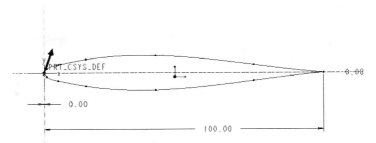

Fig. 9.15.

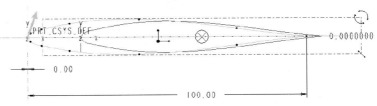

Fig. 9.16.

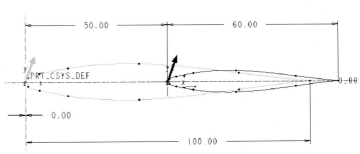

Fig. 9.17.

Fig. 9.18.

Learning Objectives:

- Learn *Sweep* and *Helical Sweep* features.

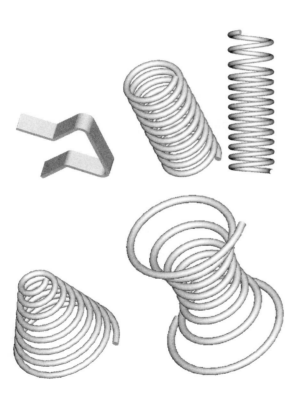

Design Information:

A spring is a flexible element used to:

- Store energy;

- Exert a force or torque over specified displacement; and

- Isolate vibrations.

Several different geometries can be used for a spring.

Helical springs are the most commonly used in engineering applications. Sometimes the pitch of the helical coils is varied to eliminate resonant surging. The primary advantage of conical springs is the nesting of coils in the fully compressed position. It results in the smallest shut height. Conical springs have a nonlinear spring rate. However, the spring rate can be made constant by adjusting the pitch.

Goal I: Create a sweep feature

1. Sketch the rough sweep profile.

2. Modify dimensions.

3. Add two fillets.

4. Create the sweep section.

5. Accept the feature creation.

Sequence of Steps

Goal II: Mirror the sweep feature.

1. Select the sweep feature.

2. Define the mirror plane.

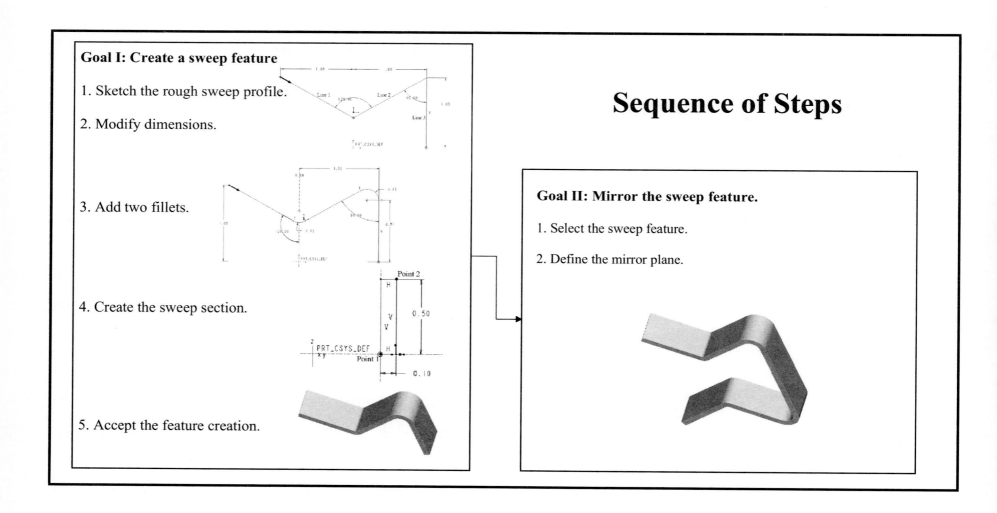

Goal	Step	Commands
Open a new file for the spring part	1. Set up the working directory.	FILE → SET WORKING DIRECTORY → *Select the working directory* → OK
	2. Open a new file.	FILE → NEW → *Part* → *Solid* → spring1 → OK
Create a sweep	3. Start "Protrusion – Sweep" feature.	Feature → Create → Solid → Protrusion → Sweep → Solid → Done
	4. Define the sketching plane.	Sketch Traj → Setup New → Plane → Pick → *Select the FRONT datum plane* → Okay → Top → *Select the TOP datum plane*
	5. Draw a rough section.	╲ → *Pick points 1, 2, 3, and 4* → *Middle Mouse* **Refer Fig. 10.1.**
	6. Modify the dimensions.	↖ → *Double click each dimension and enter the corresponding value* **Refer Fig. 10.2.**
	7. Add two fillets.	⊾ → *Select lines 1 and 2* → *Select lines 2 and 3*

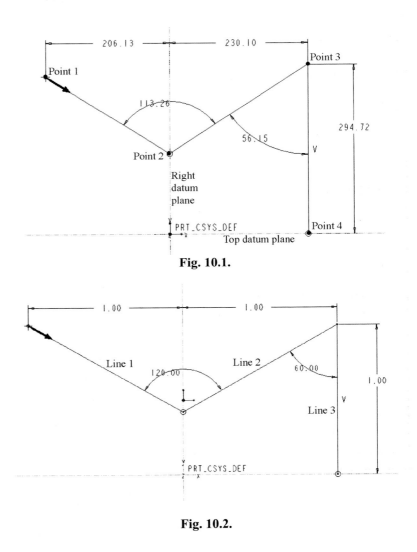

Fig. 10.1.

Fig. 10.2.

Goal	Step	Commands
Create a sweep (Continued)	8. Modify the dimensions of the fillets.	↖ → *Double click each fillet dimension and enter 0.15* **Refer Fig. 10.3.**
	9. Exit sketcher.	✔
	10. Draw the sweep section.	▢ → *Select two corners (points 1 and 2)* **Refer Fig. 10.4.**
	11. Modify the dimensions.	↖ → *Double click each dimension and enter corresponding value* **Refer Fig. 10.4.**
	12. Exit sketcher.	✔
	13. Accept the feature creation after previewing.	Preview → VIEW → DEFAULT ORIENTATION → OK **Refer Fig. 10.5.**
Mirror the protrusion feature	14. Mirror the feature.	Copy → Mirror → Select → Dependent → Done → Select → Pick → *Select the protrusion* → Done Sel → Done → Plane → Pick → *Select the TOP datum plane* **Refer Fig. 10.6.**
Save the file and exit ProE	15. Save the file and exit ProE.	FILE → SAVE → SPRING1.PRT → ☑ → FILE → EXIT → YES

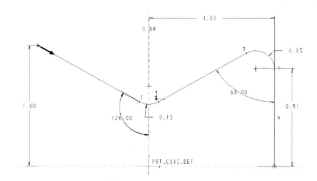

Fig. 10.3.

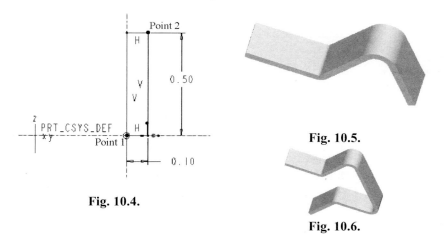

Fig. 10.4.

Fig. 10.5.

Fig. 10.6.

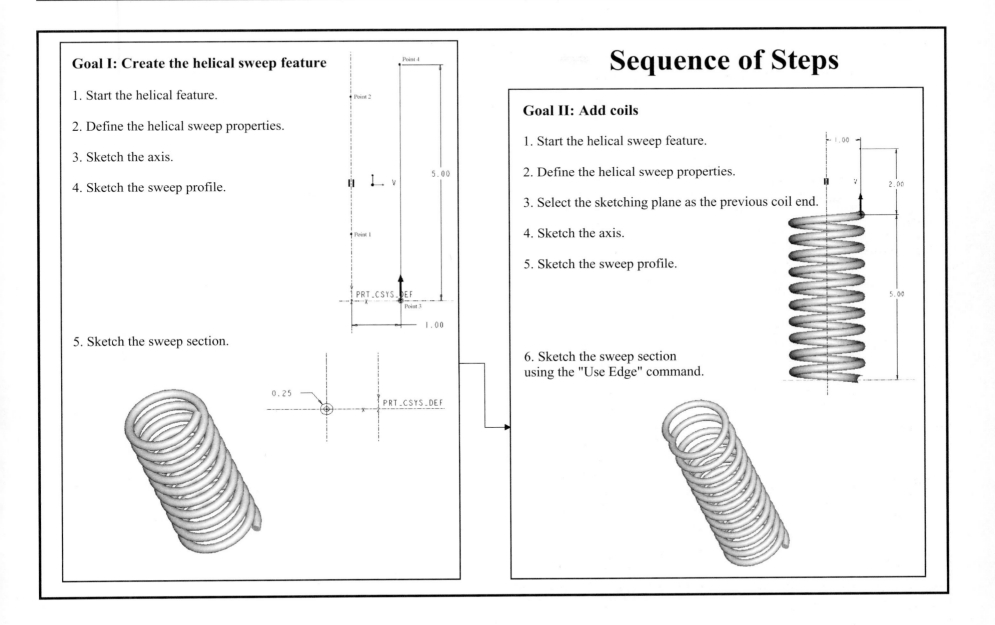

Goal I: Create the helical sweep feature

1. Start the helical feature.

2. Define the helical sweep properties.

3. Sketch the axis.

4. Sketch the sweep profile.

5. Sketch the sweep section.

Sequence of Steps

Goal II: Add coils

1. Start the helical sweep feature.

2. Define the helical sweep properties.

3. Select the sketching plane as the previous coil end.

4. Sketch the axis.

5. Sketch the sweep profile.

6. Sketch the sweep section using the "Use Edge" command.

Goal	Step	Commands
Open a new file for the spring part	1. Set up the working directory.	FILE → SET WORKING DIRECTORY → *Select the working directory* → OK
	2. Open a new file.	FILE → NEW → *Part* → *Solid* → Spring2 → OK
Create the close coiled spring	3. Start "Protrusion – Sweep" feature.	Feature → Create → Solid → Protrusion → Advanced → Solid → Done → Helical Swp → Done
	4. Define the helical sweep properties.	Constant → Thru Axis → Right Handed → Done
	5. Select the sketching plane.	Setup New → Plane → Pick *Select FRONT datum plane* → Okay
	6. Orient the sketching plane.	Right → Plane → Pick → *Select the RIGHT datum plane*
	7. Sketch the axis.	⟍ ⟍ ⋮ → ⋮ → *Pick points 1 and 2 on the RIGHT datum*
	8. Sketch the sweep profile.	⋮ ⟍ ⋮ → ⟍ → *Pick points 3 and 4* → *Middle Mouse* **Refer Fig. 10.7.**
	9. Modify the dimensions.	↖ → *Double click the height dimension* → 5 → **ENTER** → *Double click the horizontal placement dimension* → 1 → **ENTER** **Refer Fig. 10.7.**
	10. Exit sketcher.	✔

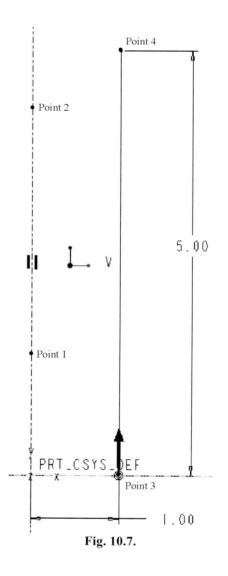

Fig. 10.7.

Goal	Step	Commands
Create the close coiled spring (Continued)	11. Enter pitch value.	0.5 → ✓ **Pitch defines the distance between two adjacent coils.**
	12. Sketch the sweep section.	○ → *Select the center of the circle at the intersection of the cross-hairs* → *Select a point to define the circle* → ↖ → *Double click the diameter dimension* → 0.25 → **ENTER** **Refer Fig. 10.8.** **The term sweep section refers to the section that is swept along the helix. The axis of the helix and the radius of revolution are defined in steps 7 and 8.**
	13. Exit sketcher.	✔
	14. Accept the feature creation after previewing.	Preview → VIEW → DEFAULT ORIENTATION → OK **Refer Fig. 10.9.**
Add coils to the spring	15. Start "Protrusion – Sweep" feature.	Create → Solid → Protrusion → Advanced → Solid → Done → Helical Swp → Done
	16. Define the helical sweep properties.	Constant → Thru Axis → Right Handed → Done

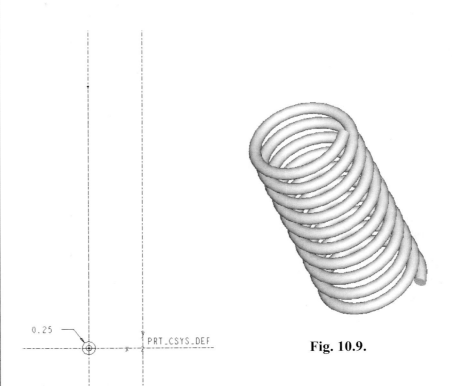

0.25

PRT_CSYS_DEF

Fig. 10.8.

Fig. 10.9.

Goal	Step	Commands
Add coils to the spring (Continued)	17. Select the sketching plane.	Setup New → Plane → Query Sel → *Select the flat face of the spring* → Accept → Flip → Okay **Refer Figs. 10.10 and 10.11.**
	18. Orient the sketching plane.	Top→ Plane → Pick → *Select the TOP datum plane*
	19. Select references.	*Select the TOP and RIGHT datum planes as references* **Refer Fig. 10.12.**
	20. Sketch the axis.	↘ ↘ → ⋮ → *Pick points 1 and 2 on the RIGHT datum* **Refer Fig. 10.13.**
	21. Sketch the sweep profile.	⋮ ↘ → ↘ → *Pick points 3 and 4* → *Middle Mouse* **IF ProE prompts: Should the selected entities be aligned, then, select NO.** **Refer Fig. 10.13.**
	22. Modify the dimensions.	↖ → *Double click each dimension and enter the corresponding values* **Refer Fig. 10.13.**
	23. Exit sketcher.	✔

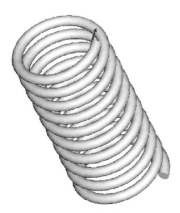

Fig. 10.10.

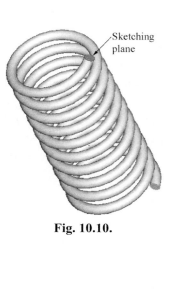

Fig. 10.11.

Fig. 10.12.

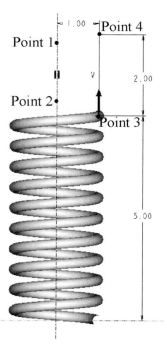

Fig. 10.13.

Goal	Step	Commands
Add coils to the spring (Continued)	24. Enter pitch value.	<u>0.75</u> → ✓
	25. Sketch the sweep section.	○ → *Select the center of the circle at the intersection of the cross-lines → Select a point to define the circle →* ↖ *→ Double click on the diameter dimension →* 0.25 *→* ✓ **Refer Fig. 10.14.**
	26. Exit sketcher.	✓
	27. Accept the feature creation after previewing.	Preview → VIEW → DEFAULT ORIENTATION → OK **Refer Fig. 10.15.**
Save the file and exit ProE	28. Save the file and exit ProE.	FILE → SAVE → SPRING2.PRT → ✓ → FILE → EXIT → YES

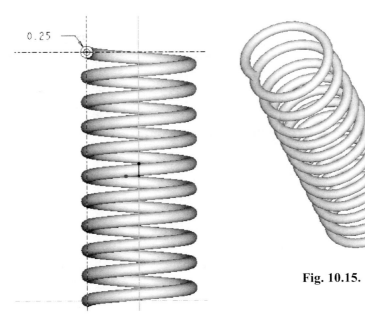

0.25

Fig. 10.14.

Fig. 10.15.

Sequence of Steps

Goal I: Create the helical sweep feature

1. Start the helical feature.

2. Define the helical sweep properties.

3. Sketch the axis.

4. Sketch the sweep profile.

5. Sketch the sweep section.

6. Accept the feature creation.

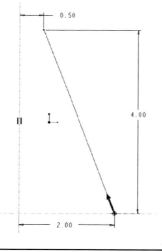

Goal	Step	Commands
Open a new file for the spring part	1. Set up the working directory.	FILE → SET WORKING DIRECTORY → *Select the working directory* → OK
	2. Open a new file.	FILE → NEW → *Part* → *Solid* → conicalspring → OK
Create the conical spring	3. Start "Protrusion – Sweep" feature.	Feature → Create → Solid → Protrusion → Advanced→ Solid → Done → Helical Swp → Done
	4. Define the helical sweep properties.	Constant → Thru Axis → Right Handed → Done
	5. Select the sketching plane.	Setup New → Plane → Pick *Select the FRONT datum plane* → Okay
	6. Orient the sketching plane.	Right → Plane → Pick → *Select the RIGHT datum plane*
	7. Sketch the axis.	＼ ▸ ＼ ⋮ → ⋮ → *Pick points 1 and 2 on the RIGHT datum plane*
	8. Sketch the sweep profile.	⋮ ▸ ＼ ⋮ → ＼→ *Pick points 3 and 4 → Middle Mouse* **Refer Fig. 10.16.**
	9. Add dimensions	⊢⟷⊣ → *Select point 3 and centerline → Middle Mouse → Select point 4 and centerline → Middle Mouse → Select points 3 and 4 → Middle Mouse to place the vertical dimension* **Refer Fig. 10.16.**

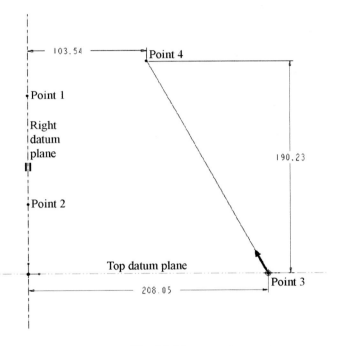

Fig. 10.16.

Goal	Step	Commands
Create the conical spring (Continued)	10. Modify the dimensions.	⬉ → *Double click each dimension and enter the corresponding value* **Refer Fig. 10.17.**
	11. Exit sketcher.	✔
	12. Enter pitch value.	0.5 → ✔
	13. Sketch the sweep section.	○ → *Select the center of the circle at the intersection of the cross-lines* → *Select a point to define the circle* → ⬉ → *Double click the diameter dimension* → 0.25 → **ENTER** **Refer Fig. 10.18.**
	14. Exit sketcher.	✔
	15. Accept the feature creation after previewing.	Preview → VIEW → DEFAULT ORIENTATION → OK **Refer Fig. 10.19.**
Save the file and exit ProE	16. Save the file and exit ProE.	FILE → SAVE → CONICALSPRING.PRT → ✔ → FILE → EXIT

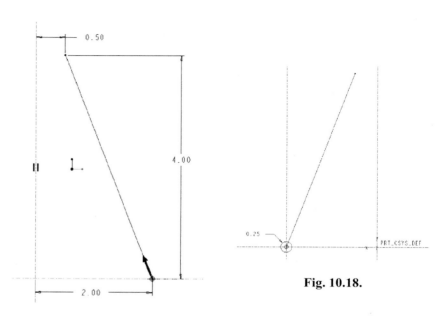

Fig. 10.17.

Fig. 10.18.

Fig. 10.19.

Sequence of Steps

Goal I: Create the helical sweep feature

1. Start the helical feature.

2. Define the helical sweep properties (variable pitch).

3. Sketch the axis.

4. Sketch the sweep profile.

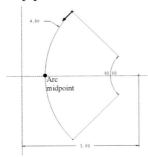

5. Divide the arc at the midpoint.

6. Define the pitch values at the intermediate locations.

7. Sketch the sweep section.

8. Accept the feature creation.

Goal	Step	Commands
Open a new file for the spring part	1. Set up working directory.	FILE → WORKING DIRECTORY → *Select working directory* → OK
	2. Open a new file.	FILE → NEW → *Part* → *Solid* → hourglassspring → OK
Create the conical spring	3. Start "Protrusion – Sweep" feature.	Feature → Create → Solid → Protrusion → Advanced→ Solid → Done → Helical Swp → Done
	4. Define the helical sweep properties.	Variable → Thru Axis → Right Handed → Done
	5. Select the sketching plane.	Setup New → Plane → Pick *Select the FRONT datum plane →* Okay
	6. Orient the sketching plane.	Right → Plane → Pick → *Select the RIGHT datum plane*
	7. Sketch the axis.	↘ · ↘ ⋮ → ⋮ → *Pick points 1 and 2 on the RIGHT datum →* Pick points 3 and 4 on the TOP *datum plane* **Refer Fig. 10.20.**
	8. Sketch the sweep profile.	↷ · ↷ ⌇ ↷ ⌒ → ↷ → *Pick points 5 (Center of the arc),* 6 and 7 (start and end points of the arc) → *Middle Mouse to discontinue the line creation* **Refer Fig. 10.20.**

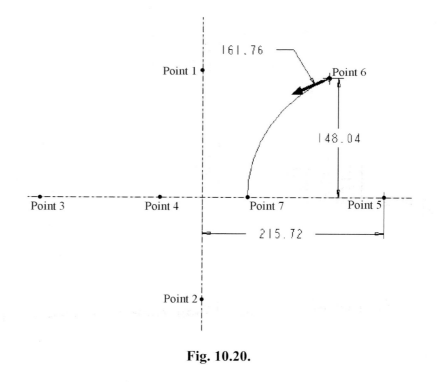

Fig. 10.20.

Goal	Step	Commands
Create the conical spring (Continued)	9. Dimension the arc.	|↔| → *Select the two end points of the arc (points 6 and 7), then the arc → Middle Mouse to place the angular dimension* **Refer Fig. 10.21.**
	10. Modify the dimensions.	↖ → *Double click each dimension and enter corresponding value (Start with the placement dimensions and then the radius)* **Refer Fig. 10.22.**
	11. Mirror the entities.	↖ → *Select the arc → EDIT → MIRROR → Select the horizontal centerline* **Refer Fig. 10.23.**
	12. Divide the arc.	⌐ ▸ ⌐ ⌐ ⌐ → ⌐ → *Select the intersection of the arc and the TOP datum plane* **Refer Fig. 10.23.**
	13. Exit sketcher.	✔
	14. Enter pitch value.	(Pitch value at the start) <u>1</u> → ✔ → (Pitch value at the end) <u>1</u> → ✔

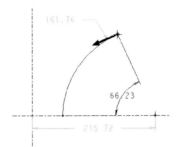

Fig. 10.21.

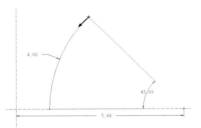

Fig. 10.22.

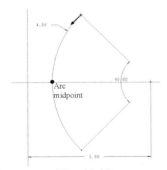

Fig. 10.23.

Goal	Step	Commands
Create the conical spring (Continued)	15. Define the pitch values at the intermediate locations.	Define → Add Point → Pick → **Refer Fig. 10.24.** *Select midpoint on the arc →* (Pitch) <u>0.5</u> → ✓ → Done Sel Done/Return → Done **Refer Fig. 10.25.**
	16. Sketch the sweep section.	O → *Select the center of the circle at the intersection of the cross-lines → Select a point to define the circle → ⬉ → Double click the diameter dimension →* <u>0.25</u> → **ENTER** **Refer Fig. 10.26.**
	17. Exit sketcher.	✔
	18. Accept the feature creation after previewing.	Preview → VIEW → DEFAULT ORIENTATION → OK **Refer Fig. 10.27.**
Save the file and exit ProE	19. Save the file and exit ProE.	FILE → SAVE → <u>HOURGLASSSPRING.PRT</u> → ✓ → FILE → EXIT → YES

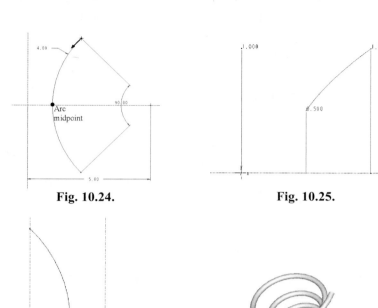

Fig. 10.24.   **Fig. 10.25.**

Fig. 10.27.

Fig. 10.26.

Learning Objectives:

- Learn *Protrusion – Variable Section Sweep*, *Cut – Variable Section Sweep* and *Datum – Graph* features.

- Practice *Protrusion – Extrude* and *Relations*, *Datum – Curves* features.

- Learn *Suppress* command.

Design Information:

Cam-follower systems are commonly used for precise motion generation. The term "axial cam" refers to cams whose follower motion is in the axial direction (parallel to the axis of the cam rotation). The follower can ride either in a track or on a rib.

Cam design involves plotting the rise, dwell and fall periods of the follower as a displacement graph. During the rise and fall periods, the follower experiences motion whereas during the dwell period, there is no follower motion. The displacement graph can be constructed by connecting the desired extreme positions of the follower using straight lines.

This simplistic approach leads to infinite acceleration and jerk and therefore, is highly undesirable. This lesson shows how to create this naive axial cam. However, the same procedure is used to create sophisticated cams.

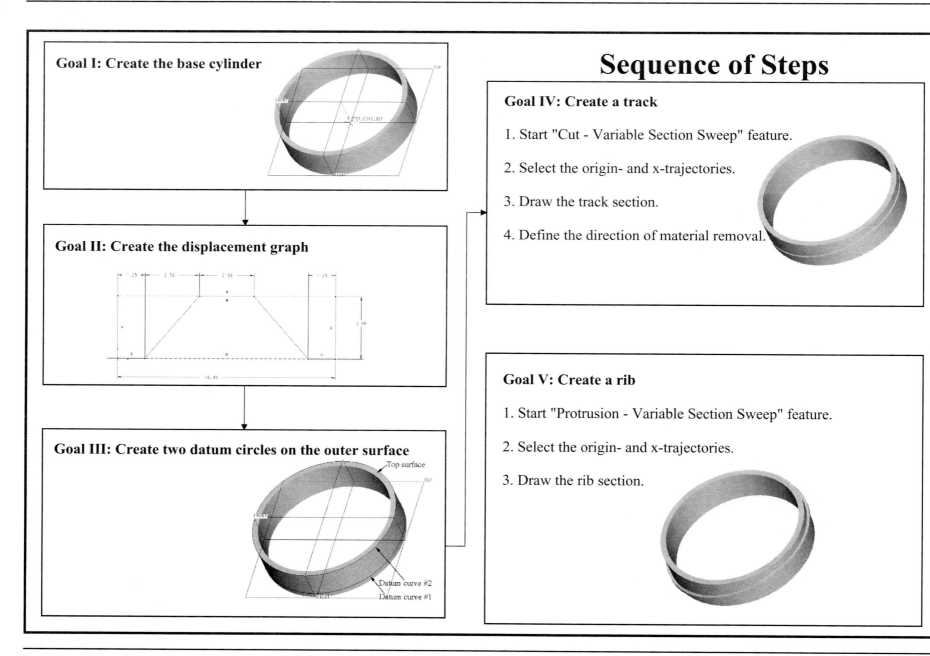

Goal I: Create the base cylinder

Sequence of Steps

Goal IV: Create a track

1. Start "Cut - Variable Section Sweep" feature.

2. Select the origin- and x-trajectories.

3. Draw the track section.

4. Define the direction of material removal.

Goal II: Create the displacement graph

Goal III: Create two datum circles on the outer surface

Top surface

Datum curve #2

Datum curve #1

Goal V: Create a rib

1. Start "Protrusion - Variable Section Sweep" feature.

2. Select the origin- and x-trajectories.

3. Draw the rib section.

Goal	Step	Commands
Open a new file for the axial cam part	1. Set up the working directory.	FILE → SET WORKING DIRECTORY → *Select the working directory* → OK
	2. Open a new file.	FILE → NEW → *Part* → *Solid* → axialcam → OK
Create the base cylinder	3. Start "Protrusion – Extrude" feature.	Feature → Create → Solid → Protrusion → Extrude → Thin → Done
	4. Define the direction of extrusion.	One Side → Done
	5. Define the sketching plane.	Setup New → Plane → Pick *Select the TOP datum plane* → Okay
	6. Orient the sketching plane.	Right → Plane → Pick → *Select the RIGHT datum plane*
	7. Draw a circle.	O → *Select the center of the circle at the intersection of the FRONT and RIGHT planes* → *Select a point to define the outer edge of the circle* **Refer Fig. 11.1.**
	8. Modify the diameter.	↖ → *Double click the diameter dimension* → 48 → **ENTER**
	9. Exit sketcher.	✔
	10. Define the direction and thickness of material creation.	Flip → Okay → 2 → ✔ **The arrow should point away from the center.**
	11. Define the depth.	Blind → Done → 12 → ✔

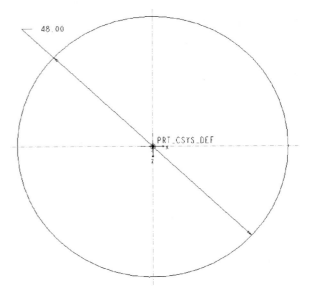

48.00

PRT_CSYS_DEF

Fig. 11.1.

Goal	Step	Commands
Create the base cylinder (Continued)	12. Accept the feature creation.	OK → VIEW → DEFAULT ORIENTATION **Refer Fig. 11.2.**
Create datum graph	13. Start "Datum – Graph" feature.	INSERT → DATUM → GRAPH → Profile → ✓ **ProE opens the sketcher.**
	14. Define a coordinate system.	⤴ → *Select a point in the graphics window*
	15. Draw a rectangle.	▢ → *Pick points 1 and 2* **Refer Fig. 11.3.**
	16. Modify the dimensions.	↖ → *Double click the horizontal dimension* → 10 → ***ENTER*** → *Double click the height dimension* → 3→ ***ENTER*** **Refer Fig. 11.3.**
	17. Change the rectangle into a construction entity.	↖ → *Select the four sides* (Hold "Shift" key while selecting multiple entities) → EDIT → TOGGLE CONSTRUCTION **Refer Fig. 11.3.**

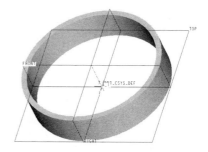

Fig. 11.2.

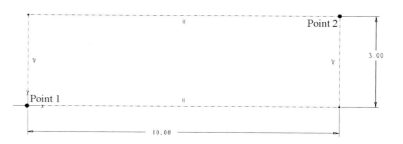

Fig. 11.3.

Goal	Step	Commands
Create datum graph (Continued)	18. Create the displacement profile.	＼ → *Pick points 3, 4, 5, 6, 7 and 8 → Middle Mouse* **Refer Fig. 11.4.**
	19. Add dimensions.	↦ → *Click on Points 4 and 5 → Middle Mouse to place the horizontal dimension*
	20. Modify dimensions.	↖ → *Double click each dimension and enter the corresponding values* **Refer Fig. 11.5.**
	21. Exit sketcher.	✔
Create datum curves	22. Start "Datum – Curve" command.	INSERT → DATUM → CURVE → Sketch → Done
	23. Set up a sketching plane that is 2″ above the bottom surface of the cylinder.	Setup New → Make Datum → Offset → Plane → Pick → *Select the TOP datum* → Enter Value → 2 → ✔ → Done → Okay
	24. Orient the sketching plane.	Right → Plane → Pick → *Select the RIGHT datum plane*
	25. Create a datum circle.	□ → *Single (in the Type window) → Select the top half of the outer circle →Select the bottom half of the outer circle* **Refer Fig. 11.6.**
	26. Exit sketcher.	✔
	27. Finish the datum curve.	OK → VIEW → DEFAULT ORIENTATION

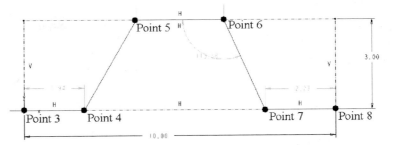

Fig. 11.4.

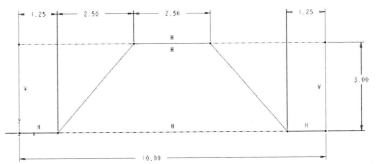

Fig. 11.5.

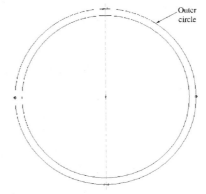

Fig. 11.6.

Goal	Step	Commands
Create datum curves (Continued)	28. Start "Datum – Curve" command.	INSERT → DATUM → CURVE → Sketch → Done
	29. Setup a sketching plane.	Setup New → Plane → Pick → *Select the top surface of the cylinder* → Okay **Refer Fig. 11.7.**
	30. Orient the sketching plane.	Right → Plane → Pick → *Select the RIGHT datum plane*
	31. Create a datum circle.	□ → *Select the top half of the outer circle* → *Select the bottom half of the outer circle*
	32. Exit sketcher.	✔ **Refer Fig. 11.7.**
	33. Finish the datum curve.	OK → VIEW → DEFAULT ORIENTATION
Create a track	34. Start "Cut – Variable Section Sweep" feature.	Create → Solid → Cut → Advanced → Solid → Done → Var Sec Swp → Done → Nrm to Origin Traj → Done →
	35. Select the origin- and x-trajectories.	Select traj (Origin Traj) → One by one → Select → Pick → *Select the two halves of the datum curve #1* → Done → Select traj (X-Traj) → One by one → Select → Pick → *Select the two halves of the datum curve #2* → Done→ Done

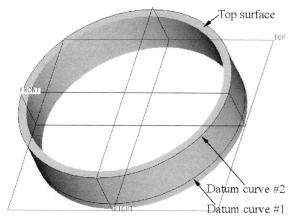

Fig. 11.7.

Origin Trajectory – The origin of the swept section (cross-hairs) is located on the origin trajectory. The option "Normal to origin trajectory" specifies that the section is always normal to the origin trajectory.

X-Trajectory – The positive x-axis of the swept section's coordinate system points towards the x-trajectory.

Goal	Step	Commands
Create a track (Continued)	36. Draw a rectangle.	▢ → *Pick points 1 and 2* **Refer Fig. 11.8.** **If ProE prompts whether to align the highlighted points with the inner edge of the cylinder, then select NO.**
	37. Modify the dimensions of the rectangle.	↖ → *Double click the two side dimensions and enter 1* **Refer Fig. 11.8.**
	38. Add relationship.	SKETCH → RELATIONS → Add → sd7 = evalgraph ("profile",trajpar*10) * 2 → ✓ → ✓ **Refer Fig. 11.8(b).** **Sd7 (the distance of the section from the origin) may be slightly different in your model.**
	39. Exit sketcher.	✔
	40. Accept the direction of material removal.	Okay
	41. Finish the feature creation.	OK → VIEW → DEFAULT ORIENTATION **Refer Fig. 11.9.**
Suppress the track feature	42. Suppress the feature.	Suppress → Normal → Select → Pick → *Select the cut feature from the model tree* → Done Sel → Done

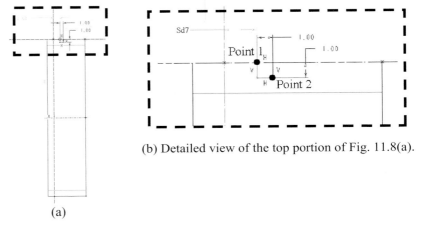

(a)

(b) Detailed view of the top portion of Fig. 11.8(a).

Fig. 11.8.

Common form of the equation is: Sd# = evalgraph("Graph Name", trajpar * Width of the graph * Horizontal Scale) * Vertical Scale
Trajpar is a normalized variable (varies between 0 and 1). If the horizontal scale is 1, then the x-axis of the graph is scaled to fit the length of the origin trajectory. The vertical scale scales the y-value of the graph.

Fig. 11.9.

Goal	Step	Commands
Create a rib	43. Start "Protrusion – Variable Section Sweep" feature.	Create → Solid → Protrusion → Advanced → Solid → Done → Var Sec Swp → Done → Nrm to Origin Traj → Done →
	44. Select the origin- and x-trajectories.	Select traj → One by one → Select → Pick → *Select the two halves of the datum curve #1* → Done → Select traj → One by one → Select → Pick → *Select the two halves of the datum curve #2* → Done → Done
	45. Draw a rectangle.	▢ → *Pick Points 1 and 2* **Refer Fig. 11.10.**
	46. Modify the dimensions of the rectangle.	↖ → *Double click the dimensions of the rectangle and enter 1* **Refer Fig. 11.10.**
	47. Add relationship.	SKETCH → RELATION → Add → <u>sd8 = evalgraph ("profile",trajpar*10) * 2</u> → ✔ → ✔
	48. Exit sketcher.	✔

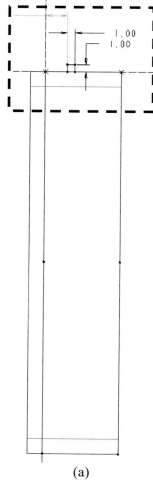

(a)

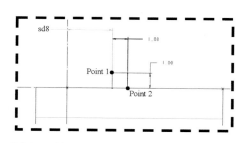

(b) Detailed view of the top portion of Fig. 11.10(a).

Fig. 11.10.

Goal	Step	Commands
Create a rib (Continued)	49. Finish the feature creation.	OK → VIEW → DEFAULT ORIENTATION **Refer Fig. 11.11.**
	50. Turn off datum curves.	VIEW → DISPLAY SETTINGS → MODEL DISPLAY → *Click on the "Shade" tab* → Unselect "With datum curves" → OK
Save the file and exit ProE	51. Save the file and exit ProE.	FILE → SAVE → AXIALCAM.PRT → ✔ → FILE → EXIT → Yes

Fig. 11.11.

LESSON 12
RADIAL PLATE CAM

Learning Objectives:

- Learn **Datum – Curves – From Equations**, **Sketched Hole** and **Pattern – Radial** features.

- Practice **Protrusion – Extrude, Straight Hole** and **Datum – Curves – Sketch** features.

- Learn **Suppress** and **Resume** commands.

Design Information:

Radial cams move the follower in the radial direction. In a radial plate cam, an external force is required to keep the contact between the follower and the cam. Therefore, these open cams are known as force closed cams.

The cam profile is designed to minimize both jerk and peak velocity. Some of the commonly used profiles are modified trapezoidal acceleration profile, modified sinusoidal acceleration profile and cycloidal displacement profile.

As the cycloidal displacement profile creates the least amount of jerk, it is commonly used in cam design. While we create a cycloidal plate cam in this lesson, the procedure for creating other cam profiles from equations is identical.

Sequence of Steps

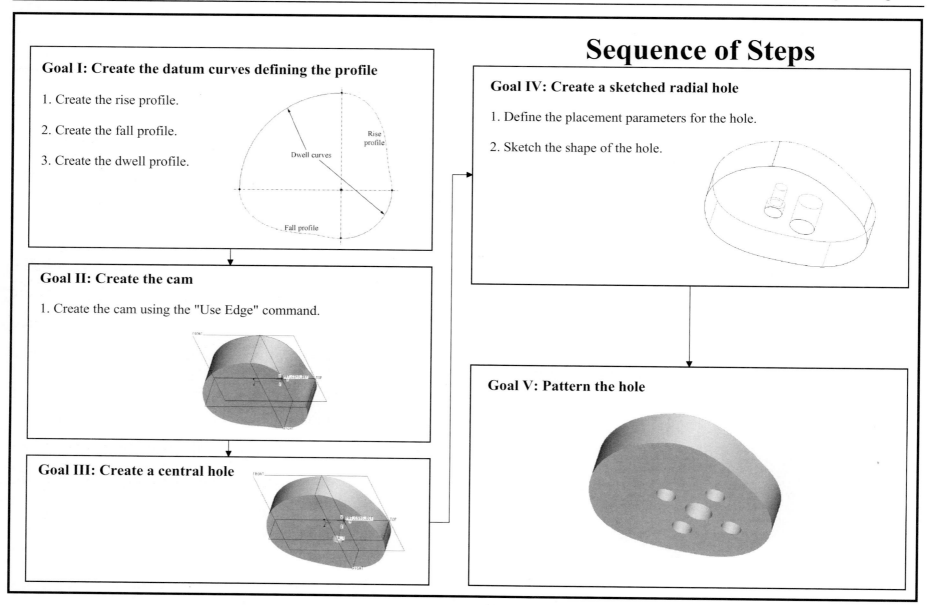

Goal I: Create the datum curves defining the profile

1. Create the rise profile.

2. Create the fall profile.

3. Create the dwell profile.

Goal II: Create the cam

1. Create the cam using the "Use Edge" command.

Goal III: Create a central hole

Goal IV: Create a sketched radial hole

1. Define the placement parameters for the hole.

2. Sketch the shape of the hole.

Goal V: Pattern the hole

Goal	Step	Commands
Open a new file for the bearing part	1. Set up the working directory.	FILE → SET WORKING DIRECTORY → *Select the working directory* → OK
	2. Open a new file.	FILE → NEW → *Part* → *Solid* → platecam → OK
Create a datum curve for the rise	3. Start creating the datum curve.	INSERT → DATUM → CURVE → From Equation → Done
	4. Select the coordinate system.	Select → Pick → *Select the default coordinate system PRT_CSYS_DEF* → Cylindrical **ProE opens equation editor.**
	5. Enter the equations of cycloidal rise. **Refer Fig. 12.1. for equations.**	Input the equations. **Refer Fig. 12.2.**
	6. Exit equation editor.	**In the equation editor window:** FILE → SAVE → FILE → EXIT

The equation for cycloidal rise is:

$$r = r_i + \frac{h}{2}\left\{\left[1 - \cos\left(180\frac{\theta}{\beta}\right)\right] - \frac{1}{4}\left[1 - \cos\left(360\frac{\theta}{\beta}\right)\right]\right\}$$

where

r is the radius at angle θ.

r_i is the base radius. Assumed to be 1″ for this design task.

h is the rise. Assumed to be 1″.

β is the angle during which the rise occurs. Assumed to be 90^0.

The equations should be written in parametric form in terms of variable t. ProE automatically varies t between 0 and 1 and evaluates the value of r, θ and z. As θ varies between 0 and 90, we can write θ as:

$$\theta = 90 \times t$$

By rearranging the terms in the equation for r, we get

$$r = r_i + \frac{h}{2}\left\{\left[1 - \cos(180t)\right] - \frac{1}{4}\left[1 - \cos(360t)\right]\right\}$$

Fig. 12.1.

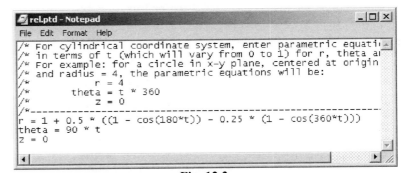

Fig. 12.2.

Goal	Step	Commands
Create a datum curve for the rise (Continued)	7. Accept the datum curve.	Preview → OK **Refer Fig. 12.3.**
Create a datum for the fall	8. Start creating the datum curve.	INSERT → DATUM → CURVE → From Equation → Done
	9. Select the coordinate system.	Select → Pick → *Select the default coordinate system PRT_CSYS_DEF* → Cylindrical **ProE opens equation editor.**
	10. Enter the equations of cycloidal fall. **Refer Fig. 12.4. for the equations.**	Input the equations. **Note that the fall occurs between 180^0 and 270^0.** **Refer Fig. 12.5.**
	11. Exit equation editor.	**In the equation editor window:** FILE → SAVE → FILE → EXIT

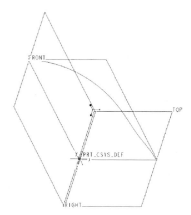

Fig. 12.3.

The equation for cycloidal fall is:

$$r = r_i + \frac{h}{2}\left\{\left[1 + \cos\left(180\frac{\theta}{\beta}\right)\right] - \frac{1}{4}\left[1 - \cos\left(360\frac{\theta}{\beta}\right)\right]\right\}$$

Fig. 12.4.

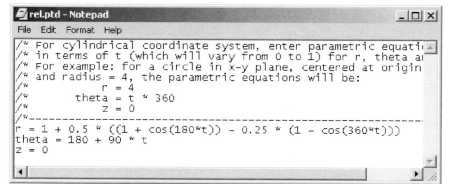

Fig. 12.5.

Goal	Step	Commands
Create a datum for the fall (Continued)	12. Accept the datum curve.	Preview → OK **Refer Fig. 12.6.**
Create datum curves for the dwell	13. Start creating the dwell datum curves.	INSERT → DATUM → CURVE → Sketch → Done
	14. Select the sketching plane.	Setup New → Plane → Pick → *Select the FRONT datum plane* → Okay
	15. Orient the sketching plane.	Right → *Select the RIGHT datum plane*
	16. Add references.	**If the references window is not visible, then activate it using SKETCH → REFERENCES command.** *Select datum curves (rise and fall)* **Refer Fig. 12.7.**
	17. Draw the two dwell curves.	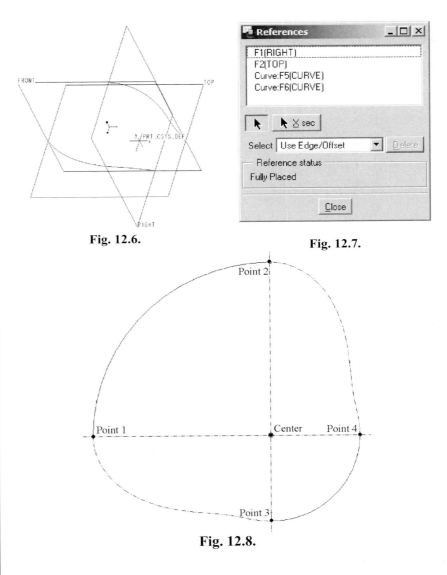 → *Select the center and points 1 and 2 → Select the center and points 3 and 4* **Make sure that points 1, 2, 3 and 4 are aligned with the corresponding points on the datum curves and datum planes.** **Refer Fig. 12.8.**
	18. Exit sketcher.	✔

Fig. 12.6.

References

F1(RIGHT)
F2(TOP)
Curve:F5(CURVE)
Curve:F6(CURVE)

Select [Use Edge/Offset ▼] Delete

Reference status
Fully Placed

Close

Fig. 12.7.

Point 2

Point 1   Center   Point 4

Point 3

Fig. 12.8.

Goal	Step	Commands
Create datum curves for the dwell (Continued)	19. Accept the feature creation after previewing.	Preview → VIEW → DEFAULT ORIENTATION → OK
Create the cam	20. Start "Protrusion – Extrude" feature.	Feature → Create → Solid → Protrusion → Extrude → Solid → Done
	21. Define the direction of extrusion.	One Side → Done
	22. Set up the sketching plane.	Use Prev
	23. Define the direction of feature creation.	Okay
	24. Select the datum curves as the edges.	SKETCH → EDGE → USE → Single → *Select the four datum curves* → Close
	25. Exit sketcher.	✔ **If ProE prompts the message "the section is not closed," then delete the dwell datum curves and repeat step 17. Points 1, 2, 3 and 4 must be aligned with the corresponding points on the fall and rise datum curves.**
	26. Define the depth.	Blind → Done → 1 → ✔
	27. Accept the feature creation after previewing.	Preview → VIEW → DEFAULT ORIENTATION → OK **Refer Fig. 12.9.**

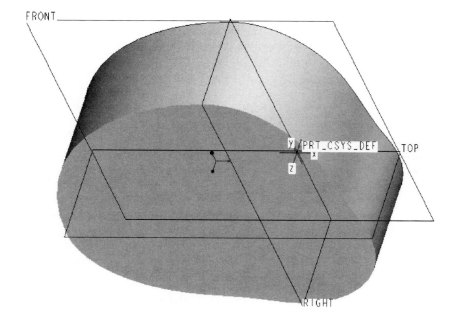

FRONT

PRT_CSYS_DEF

TOP

RIGHT

Fig. 12.9.

Goal	Step	Commands
Create a hole	28. Create a hole at the center.	Create→ Solid→ Hole **ProE opens the "hole" window.** **Refer Fig. 12.10.** (Hole Type) Straight → (Diameter) 0.5 → (Depth One) *Thru All* → (Primary Reference) ↖ → Query Selec → *Select the flat surface of the cam (parallel to the FRONT datum plane* → Accept → (Linear Reference) ↖ → Query Selec → *Select the TOP datum plane* → Accept → (Distance) 0 → **_ENTER_** → (Linear Reference) ↖ → Query Selec → *Select the RIGHT datum plane* → Accept → (Distance) 0 → **_ENTER_** → ✓

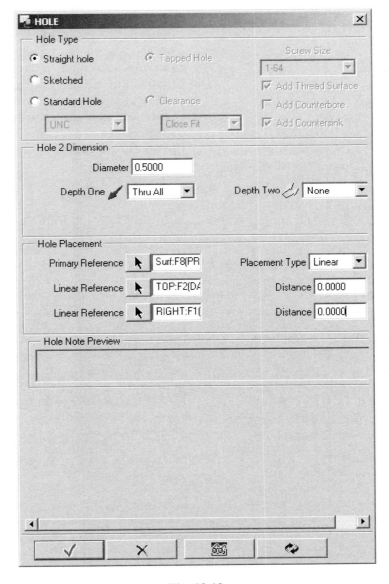

Fig. 12.10.

Goal	Step	Commands
Suppress protrusion and hole features	29. Suppress protrusion and hole features.	**As there is not enough space between the hole and outer edge of the cam, the base radius of the cam must be increased. This involves modifying one datum curve at a time. To prevent feature failure errors, we must suppress dwell curves, protrusion and hole features.** *Select the dwell curves, protrusion and hole in the model tree* (Hold shift key while selecting multiple items) → *Right Mouse → Suppress →* OK **Refer Fig. 12.11.**
Modify the datum curves	30. Redefine the rise curve.	*Select the first datum curve → Right Mouse → Redefine →* **Refer Fig. 12.12.** *Select equation from the CURVE window →* Define *→* **Refer Fig. 12.13.** Change the value of base radius from 1 to 2 → FILE → SAVE → FILE → EXIT → OK **Refer Fig. 12.14.**

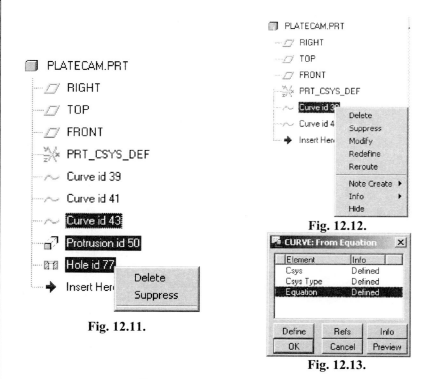

Fig. 12.11.

Fig. 12.12.

Fig. 12.13.

Fig.12.14.

Goal	Step	Commands
Modify the datum curves (Continued)	31. Redefine the fall datum curve.	*Select the second datum curve →* *Right Mouse → Redefine →* *Select equation from the CURVE window →* Define *→* Change the value of base radius from 1 to 2 *→* FILE → SAVE → FILE → EXIT → OK **Refer Fig. 12.15.**
Resume suppressed features	32. Resume the protrusion and hole features.	Feature → Resume → All → Done → Done **Refer Fig. 12.16.**
	33. Modify the dwell datum curves.	**Optional step:** **Dwell datum curves automatically align themselves with the first two datum curves. Otherwise, modify the radius of the datum curves to 2.**
Modify the hole size	34. Increase the hole size.	Modify → Value → Pick → *Select the hole from the model tree →* *Select the diameter →* 0.75 *→* ✓ → Regenerate
Create a radial hole	35. Start a sketched hole.	Feature → Create→ Solid→ Hole (Hole Type) Sketched **ProE opens sketcher.**

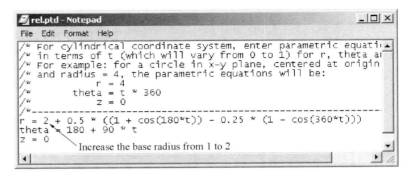

```
/* For cylindrical coordinate system, enter parametric equati
/* in terms of t (which will vary from 0 to 1) for r, theta a
/* For example: for a circle in x-y plane, centered at origin
/* and radius = 4, the parametric equations will be:
/*           r = 4
/*           theta = t * 360
/*           z = 0
/*-----------------------------------------------
r = 2 + 0.5 * ((1 + cos(180*t)) - 0.25 * (1 - cos(360*t)))
theta = 180 + 90 * t
z = 0
                      Increase the base radius from 1 to 2
```

Fig. 12.15.

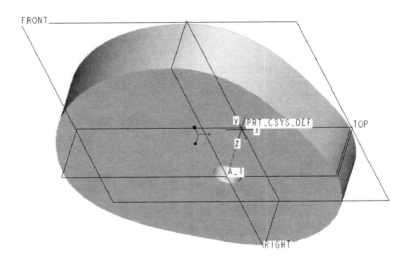

Fig. 12.16.

Goal	Step	Commands
Create the base cylinder (Continued)	9. Draw a circle.	⊙ → *Select the center of the circle as the intersection of TOP and RIGHT datum planes →* *Select a point to define the outer edge of the circle* **Refer Fig. 13.3.**
	10. Modify the diameter.	↖ → *Double click the diameter dimension →* 8→ ***ENTER***
	11. Exit sketcher.	✔
	12. Define the depth.	Blind → Done → 1 → ✔
	13. Accept the feature creation after previewing.	Preview→ VIEW → DEFAULT ORIENTATION → OK **Refer Fig. 13.4.**

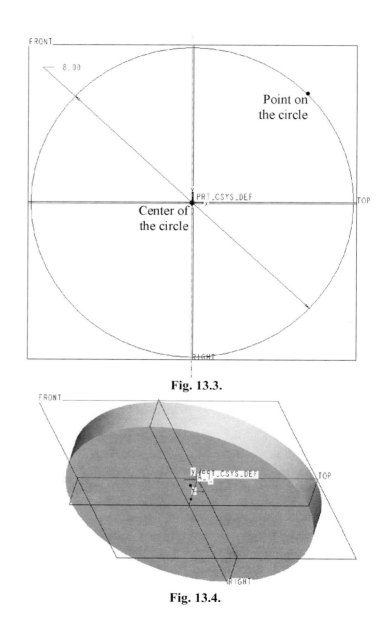

Fig. 13.3.

Fig. 13.4.

Goal	Step	Commands
Create the groove for the cam follower	14. Start "Cut – Extrude" feature.	Create → Solid → Cut → Extrude → Thin→ Done
	15. Define the number of sides.	One Side → Done
	16. Select the sketching plane.	Setup New → Plane → Pick → *Select the front surface of the cam (that is away from the FRONT datum)* **Refer Fig. 13.5.**
	17. Define the direction of cut.	Okay **Make sure that the arrow points into the part.**
	18. Orient the sketching plane.	TOP → *Select the TOP datum plane*
	19. Define a coordinate system.	⊥ → *Select the intersection of RIGHT and TOP datum surfaces*
	20. Create a spline.	∿ → *Pick points 1, 2, 3, 4 and 1* **Refer Fig. 13.6.**
	21. Select polar coordinate system.	↖ → *Select the spline →* ▱ *→ Click on the Coordinates tab in "Modify Spline" window → Select polar coordinate system* **Refer Fig. 13.7.**

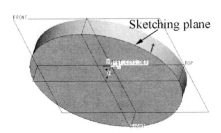

Sketching plane

Fig. 13.5.

Point 2
Point 3
Point 1
Point 4

Fig. 13.6.

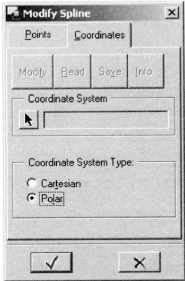

Fig. 13.7.

Goal	Step	Commands
Create the groove for the cam follower (Continued)	22. Assign the spline to the local coordinate system.	▸ *in the "Modify Spline" window → Select the coordinate system created in step 19* **The message window should display: Spline is dimensioned to the local coordinate system.** **Refer Fig. 13.8.**
	23. Read data points.	Read → **If ProE prompts: Cannot modify spline with dimensions to internal points. Delete dimensions? Click on YES** *Select "PROFILE.PTS" →* OPEN → ✓ **Refer Fig. 13.9.**
	24. Exit sketcher.	✓
	25. Define the direction of material removal.	Flip → Okay **The arrow should be pointing away from the center.** **Refer Fig. 13.9.**
	26. Specify the width of thin feature.	0.5 → ✓
	27. Define the depth.	Blind → Done → 0.5 → ✓
	28. Accept the feature creation after previewing.	Preview → VIEW → DEFAULT ORIENTATION → OK **Refer Fig. 13.10.**

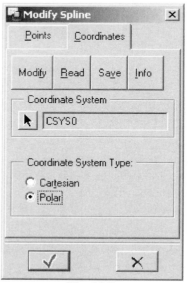

Fig. 13.8.

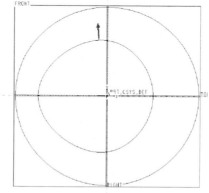

Fig. 13.9.

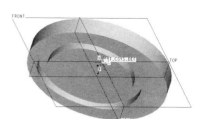

Fig. 13.10.

Goal	Step	Commands
	29. Open facecam file.	FILE → OPEN → *Select "platecam.prt"* → OPEN
	30. Create a User Defined Feature (UDF).	Feature → UDF library → Create → holepattern → ✓
	31. Define UDF options.	Stand Alone → Done (Include original part) Yes → Yes
	32. Select the features.	Add → Select → Pick → *Select the axial hole and patterned holes (the last two features in the model tree)* → Done → Done/Return
Create UDF feature	33. Define the prompts.	Single → Done/Return → the front surface → ✓ → Single → Done/Return → the TOP datum plane → ✓ → the RIGHT datum plane → ✓ → **Read the prompt and displayed reference. If it needs to be changed, click Enter prompt, or else click Next. After ensuring correct prompts, select:** Done/Return

In the stand-alone option, ProE copies the required information at the time of the UDF creation. On the other hand, the subordinate option copies the information from the original part at the run time. The second option is very useful in making sure that the holes of mating parts line up.

In the "single" option, a single prompt appears for the reference used by several features. The "multiple" option prompts references for each feature in the UDF.

Goal	Step	Commands
Create UDF feature (Continued)	34. Define the variable parameters.	*Select Var Dims in "UDF: Holepattern, Standalone" window* → Define **Refer Fig. 13.11.** Add → Select Dim → Pick → Zoom in → *Select the patterned hole depths (1.0 and 0.375 respectively.)* → The depths should be highlighted. **Refer Fig. 13.12.** Done Sel → Done/Return → Done/Return → (Enter prompts) <u>cam thickness</u> (if 1.0 is highlighted) → ✓ → <u>the depth of countersunk hole</u> (if 0.375 is highlighted) → ✓ → **If you are not sure whether the prompts correspond to their respective dimensions, click on Dim Prompts and follow instructions.** OK
	35. Save the UDF	Dbms → Save → <u>holepattern</u> → ✓ → Done-Return → Done/Return
	36. Close the window	FILE → ERASE → CURRENT → YES

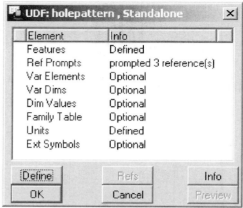

Fig. 13.11.

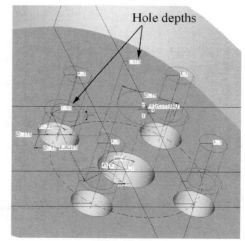

Fig. 13.12.

Goal	Step	Commands
Activate grooved cam window	37. Activate "groovedcam" window.	WINDOW → GROOVEDCAM.PRT
Insert the UDF	38. Start "holepattern" UDF.	Feature → Create → User Defined → *Select holepattern* → Open → (Retrieve reference part holepattern_gp?) Yes → Yes **ProE opens the reference part in another window. This helps in picking the corresponding datum planes.**
	39. Define the UDF options.	Independent → Done → Same Dims → Done → (Enter the cam thickness) 1 → ✓ → (Enter the countersunk hole depth) 0.25 → ✓ → Normal → Done
	40. Select references.	*Select the front surface of the cam →* **Refer Fig. 13.13.** *Select the TOP datum plane → Select the RIGHT datum plane → Click on the front surface of the cam → Done* **Refer Fig. 13.14.**
Save the file and exit ProE	41. Save the file and exit ProE.	FILE → SAVE → GROOVEDCAM.PRT → ✓ → FILE → EXIT → Yes

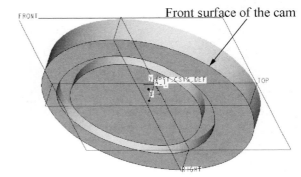

Front surface of the cam

Fig. 13.13.

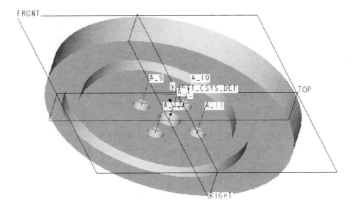

Fig. 13.14.

LESSON 14
GATEWAY ARCH

Learning Objectives:

- Learn *Swept – Blend* feature.

- Learn to *Set Up – Units*.

- Learn *PhotoRender*.

- Practice the *Use of Sections* and *Mirror* commands.

- Practice *Datum – Curves – From Equation* features.

Design Information:

Architect Eero Saarinen conceived the Gateway Arch to commemorate the westward expansion of the United States. It was completed in 1965 on the banks of the River Mississippi, in St. Louis. The stainless steel arch spans 630 ft. between its legs. At 630 ft., it is the tallest memorial in the United States. Each leg of the arch is an equilateral triangle with its sides measuring 54 ft. at the ground level and taper to 17 ft. at the top. Saarinen used an inverted catenary curve shape for the arch. Catenary curve is the shape assumed by a flexible cable hanging under its own weight between two supports. Even though it looks like a parabola, the equations are quite different.

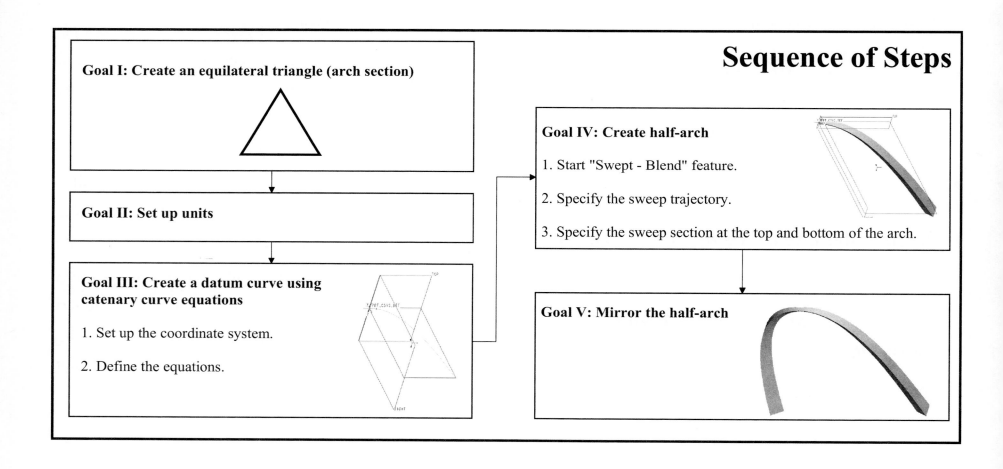

Sequence of Steps

Goal I: Create an equilateral triangle (arch section)

Goal II: Set up units

Goal III: Create a datum curve using catenary curve equations

1. Set up the coordinate system.

2. Define the equations.

Goal IV: Create half-arch

1. Start "Swept - Blend" feature.

2. Specify the sweep trajectory.

3. Specify the sweep section at the top and bottom of the arch.

Goal V: Mirror the half-arch

Goal	Step	Commands
Open a new file for the arch section	1. Set up the working directory.	FILE → SET WORKING DIRECTORY → *Select the working directory* → OK
	2. Open a new file.	FILE → NEW → *Sketch* → archsection → OK
Create the arch section (equilateral triangle)	3. Establish the reference coordinate system.	⊥ → *Click in the graphics window* **Refer Fig. 14.1.**
	4. Draw a triangle.	＼ → *Pick points 1, 2, 3 and 1 to create a triangle* → *Middle Mouse to discontinue line creation* **Refer Fig. 14.2.**

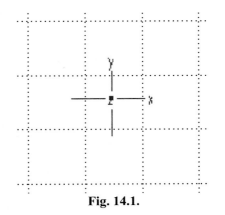

Fig. 14.1.

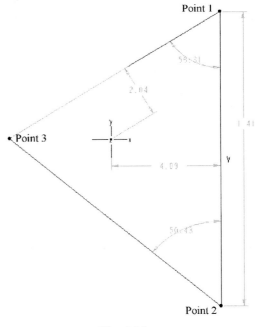

Fig. 14.2.

Goal	Step	Commands
Create arch section (Continued)	5. Dimension the triangle.	![pointer] → *Double click the angle dimension* → <u>60</u> → ***ENTER*** → *Double click the second angle dimension* → <u>60</u> → ***ENTER*** → INFO → SWITCH DIMS → *Double click the placement dimension between the reference coordinate system and the right side (sd6)* → <u>sd2/(2*sqrt(3))</u> → ***ENTER*** → Yes → YES → **Note that sd2 dimension refers to base dimension and may be different based on the order of creation.** **Refer Fig. 14.3.** ![icon] → ![arrow] → *Select the coordinate system* → *Select the apex* → *Double click the base dimension* → <u>1</u> → ***ENTER*** **Refer Fig. 14.4.**
Save the section and exit sketcher	6. Save the section.	FILE → SAVE → ARCHSECTION.SEC → ✓
	7. Exit sketcher.	FILE → CLOSE WINDOW → YES

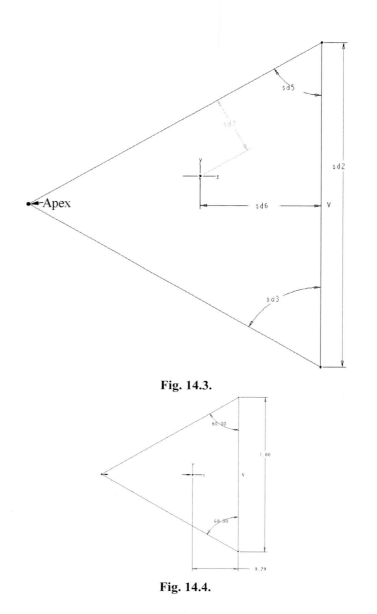

Fig. 14.3.

Fig. 14.4.

Goal	Step	Commands
Open a new file for the arch part	8. Open a new file.	FILE → NEW → Part → Solid → Arch → OK
Set up units	9. Change the units to FPS system.	Set Up → Units → *Click on Foot Pound Second (FPS)* → *Set* → OK → Close → Done
Create a datum curve	10. Establish the coordinate system.	INSERT → DATUM → CURVE → From Equation → Done → Select → Pick → *Select the default coordinate system PRT_CSYS_DEF from the model tree* → Cartesian **ProE opens the equation editor.**
	11. Input equations in the equation editor.	Input the equations **Refer Fig. 14.5.** FILE → SAVE → FILE → EXIT **ProE automatically varies t from 0 to 1.**
	12. Create the datum curve.	OK **Refer Fig. 14.6.**
Create half-arch	13. Start the swept blend feature.	Feature → Create → Solid → Protrusion → Advanced → Solid → Done → Swept Blend → Done → Sketch Sec → NrmToOriginTraj → Done
	14. Specify the sweep trajectory.	Select Traj → One by one → Select → Pick → *Select the datum curve* → Done

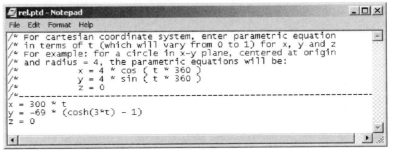

Fig. 14.5.

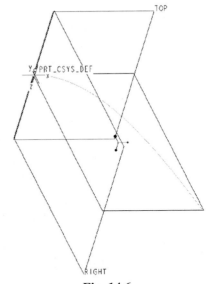

Fig. 14.6.

Goal	Step	Commands
Create half-arch (Continued)	15. Specify the sweep section orientation and the section at the top of the arch.	Automatic → Done → Okay → 0 → ✔ → SKETCH → DATA FROM FILE → *Select archsection.sec* → OPEN → (Scale) 17 → (Rotation) 0 → ✔ → ★ *Zoom in the triangle* → [⊥] → ↔ → *Select the sketch coordinate system → Select the apex → Zoom out →* [⊢⊣] → *Query select the sketcher coordinate system (CtrPnt:SectEnt on top of the PRT_CSYS_DEF) → Query select the section coordinate system (CtrPnt:SectEnt in the triangle) → Middle Mouse to place horizontal dimension → Query select the sketcher coordinate system (CtrPnt:SectEnt on top of the PRT_CSYS_DEF) → Query select the section coordinate system (CtrPnt:SectEnt in the triangle) → Middle Mouse to place vertical dimension →* [↖] → *Double click the horizontal placement dimension* → 0 → **ENTER** → *Double click the vertical placement dimension* → 0 → **ENTER** → ✔ **Refer Figs. 14.7 and 8.**

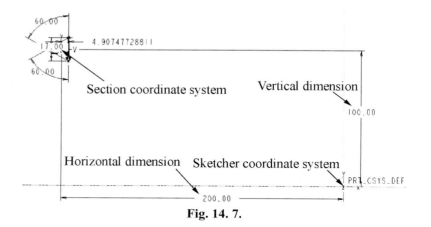

Fig. 14. 7.

The placement dimensions do not refer to the section coordinate system. Therefore, new placement dimensions (horizontal and vertical) are added between the sketcher coordinate system and the part coordinate system.

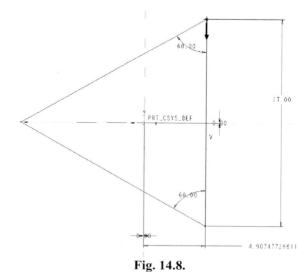

Fig. 14.8.

Goal	Step	Commands
Create half-arch (Continued)	16. Specify section orientation and section at the bottom of the arch.	Automatic → Done → Okay → 0 → ✔ → VIEW → REFIT → SKETCH → DATA FROM FILE → *Select archsection.sec* → OPEN → (Scale) 54 → (Rotation) 0 → ✔ → ★ ▣ → ↔ → *Select the sketch coordinate system* → *Select the apex* → |↔| → *Query select the sketcher coordinate system (CtrPnt:SectEnt)* ***(Refer Fig. 14.9.)*** → *Query select the section coordinate system (CtrPnt:SectEnt in the triangle)* → *Middle Mouse to place horizontal dimension* → *Query select the sketcher coordinate system (CtrPnt:SectEnt on top of the PRT_CSYS_DEF)* → *Query select the section coordinate system (CtrPnt:SectEnt in the triangle)* → *Middle Mouse to place vertical dimension* → ▸ → *Double click the horizontal placement dimension* → 0 → **ENTER** → *Double click the vertical placement dimension* → 0 → **ENTER** → ✔ **Refer Fig. 14. 10.**

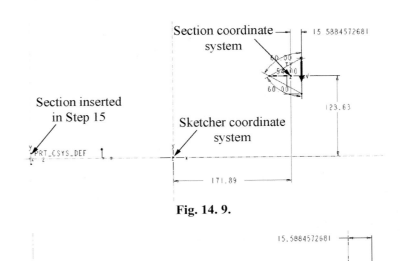

Section coordinate system

15.5884572681

123.63

Section inserted in Step 15

Sketcher coordinate system

PRT_CSYS_DEF

171.89

Fig. 14. 9.

15.5884572681

PRT_CSYS_DEF

0.00

Fig. 14.10.

Goal	Step	Commands
Create half-arch (Continued)	17. Create the arch leg.	Preview → OK → VIEW → DEFAULT ORIENTATION **Refer Fig. 14.11.**
Create the arch	18. Mirror protrusion feature.	Copy → Mirror → Select → Dependent → Done → Select → Pick → *Select the arch leg* → Done Sel → Done → Plane → Pick → *Select RIGHT datum plane* **Refer Fig. 14. 12.**
View the arch	19. View the model in the default orientation.	*Click on the following icons to switch off the datums, axes, datum points and default coordinate system.* **Refer Fig. 14.13.**
Save the part	20. Save the part.	FILE → SAVE → ARCH.PRT → ✓

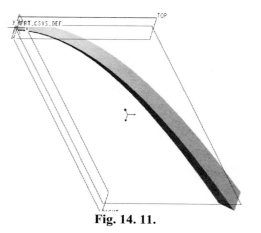

Fig. 14. 11.

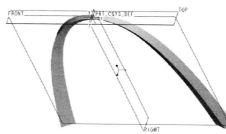

Fig. 14.12.

Fig. 14.13.

Goal	Step	Commands
Open PhotoRender	21. Open PhotoRender module.	VIEW → MODEL SETUP → PHOTORENDER **ProE opens PhotoRender menu bar.** **Refer Fig. 14.14.** **The PhotoRender options are described in Fig. 14.15.**
Render the object	22. Define the room.	⬛ → *Click on wall 2 → Select a texture for wall 2 (ANY JPEG OR GIF FILE) → Check single in the material placement window →* Assign different jpeg files for the remaining sides of the room → CLOSE
	23. Render the model.	🫖
Save the image	24. Save the image.	💾 → Arch → OK → ❌
Save the file and exit ProE	25. Save the file and exit ProE.	FILE → SAVE → ARCH.PRT → ✔ → FILE → EXIT → Yes

Fig. 14.14.

Option	Description
📂	Open an image.
💾	Save the image.
🌐	Modify rendering configuration options.
⬛	Modify room configurations. Allows the user to select different textures for the room. Also, allows the user to size the room.
💡	Modify lights.
⚪	Modify appearances.
🔲	Modify perspective views.
⬜	Modify views.
🖼	Image editor.
🫖	Render image.
✂	Exit PhotoRender.
❌	Close.

Fig. 14.15.

Lesson 15
Spur and Helical Gears

Learning Objectives:

- Learn *Set Up – Parameters* and *Set Up – Names* commands.

- Learn *Surface – New* and *Surface – Transform* commands.

- Practice *Relations*.

- Practice *Group* and *Pattern* commands.

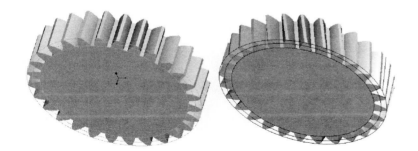

Design Information:

Gears provide an effective means for transferring power without slippage from one shaft to another. The teeth are shaped to maintain a fixed angular velocity ratio between the gears. The involute curve is the most commonly used for the tooth profile. This curve can be obtained by tracing the end of a string, as it is unwrapped from a base cylinder while keeping the string tangential to the cylinder. The pitch circle is used for calculating the velocity ratio. Addendum is the amount of tooth that sticks out above the pitch circle. The dedendum circle represents the bottom of the tooth. Typically, the dedendum circle is bigger than the base circle from which the involute profile originates. Spur gear has straight teeth. Helical gear teeth follow a helix and, therefore, share the load between multiple teeth at any instant of time.

Sequence of Steps

Goal I: Set up design parameter

1. Set up the design parameter (names and the initial values).

Goal II: Add relations

Goal III: Create datum circles

1. Create the addendum, dedendum and pitch circles.

Goal IV: Create the involute profile

1. Define the datum coordinate system.

2. Enter involute profile equations.

3. Create the pitch point.

4. Mirror the involute profile.

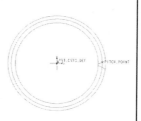

Goal V: Create a profile for cutting the teeth

1. Establish new references.

2. Sketch the profile using the existing involute profiles.

3. Trim the excess lines.

Goal VI: Create a profile for cutting the teeth

1. Start "Surface - Extrude" feature.

2. Select the cutting profile.

3. Define the depth of extrusion.

Goal VII: Create the base cylinder

1. Define the circle.

2. Define the depth of extrusion.

Goal VIII: Create the teeth

1. Copy the cutting surface.

2. Create a cut using the cutting surface.

3. Pattern the cut.

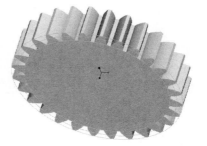

Goal	Step	Commands
Open a new file for the gear part	1. Set up the working directory.	FILE → SET WORKING DIRECTORY → *Select the working directory* → OK
	2. Open a new file.	FILE → NEW → *Part* → *Solid* → spurgear → OK
Set up design parameters	3. Set up design parameters.	Set Up → Parameters → Part → Create → Integer → N (for number of teeth) → ✓ → 28 → ✓ → Real Number → P (for diametral pitch) → ✓ → 6 → ✓ → Real Number → phi (for pressure angle) → ✓ → 25 → ✓ → Real Number → A (for addendum) → ✓ → 0 → ✓ → Real → B (for dedendum) → ✓ → 0 → ✓ → Real Number → Dp (for pitch circle diameter) → ✓ → 0 → ✓ → Real Number → Db (for base circle diameter) → ✓ → 0 → ✓ → Real Number → Dd (for dedendum circle diameter) → ✓ → 0 → ✓ → Real Number → Da (for addendum circle diameter) → ✓ → 0 → ✓ → Real Number → F (for Face width) → ✓ → 1 → ✓ → ▶ MODEL PARAMS → Info → CLOSE → Done/Return → Done Refer Fig. 15.1.

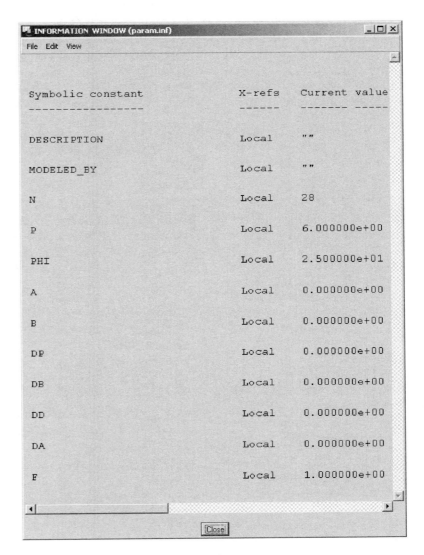

Fig. 15.1.

Goal	Step	Commands
Add relations	4. Add relations.	Relations → Part Rel → Edit Rel → Type the relations as shown in Fig. 15.2. FILE → SAVE → FILE → EXIT → Sort Rel → Show Rel → Notice the new values for different parameters → Close → Done
Create the datum circles	5. Create a datum axis.	INSERT → DATUM → AXIS → Two Planes → Plane → Pick → *Select the TOP and RIGHT datum planes*
	6. Create the addendum datum circle.	INSERT → DATUM → CURVE → Sketch → Done → Setup New → Plane → Pick → *Select the FRONT datum plane* → Okay → Top → Plane → Pick → *Select the TOP datum plane* → O → *Select the intersection of the RIGHT and TOP datum planes* → *Select a point to define the circle* → ⬆ → *Double click the diameter dimension* → Da → ***ENTER*** → Yes → Yes → ✔ → OK → VIEW → DEFAULT ORIENTATION **Refer Fig. 15.3.**

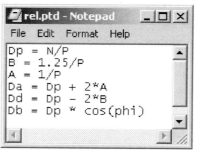

Fig. 15.2.

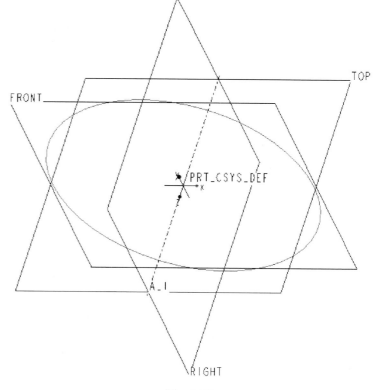

Fig. 15.3.

Goal	Step	Commands
Create the datum circles (Continued)	7. Create the dedendum datum circle.	INSERT → DATUM → CURVE → Sketch → Done → Use Prev → Okay → ⟲ → *Select the intersection of the RIGHT and TOP datum planes → Select a point to define the circle →* ↖ *→ Double click the diameter dimension →* Dd → **_ENTER_** → Yes → Yes → ✔ → VIEW → DEFAULT ORIENTATION **Refer Fig. 15.4.**
	8. Create the pitch datum circle.	INSERT → DATUM → CURVE → Sketch → Done → Use Prev → Okay → ⟲ → *Select the intersection of the RIGHT and TOP datum planes → Select a point to define the circle →* ↖ *→ Double click the diameter dimension →* Dp → **_ENTER_** → Yes → Yes → ✔ → OK → VIEW → DEFAULT ORIENTATION **Refer Fig. 15.5.**
	9. Name the circles.	Set Up → Name → Feature → Pick → *Select each circle in the model tree and enter the corresponding name* **(Refer Fig. 15.6.)** → Done Sel → Done

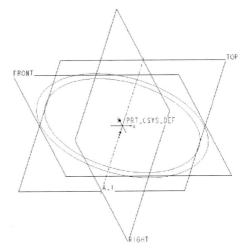

Fig. 15.4.

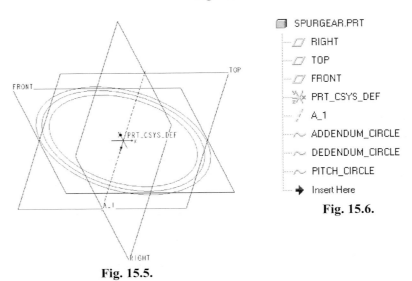

Fig. 15.5.

SPURGEAR.PRT
  RIGHT
  TOP
  FRONT
  PRT_CSYS_DEF
  A_1
  ADDENDUM_CIRCLE
  DEDENDUM_CIRCLE
  PITCH_CIRCLE
  Insert Here

Fig. 15.6.

Goal	Step	Commands
Create an involute profile datum curve	10. Start "Datum – Curve – From Equation" command.	INSERT → DATUM → CURVE → From equation → Done
	11. Define the coordinate system.	Select → Pick → *Select the default coordinate system PRT_CSYS_DEF from the model tree* → Cylindrical
	12. Enter the equations for the involute profile.	Type the equations in the equation editor **(Refer Fig. 15.7.)** FILE → SAVE → FILE → EXIT
	13. Accept the feature creation after previewing.	Preview → VIEW → DEFAULT → OK **Refer Fig. 15.8.**
Create a datum through the pitch point	14. Define the pitch point.	Insert → DATUM → POINT → Add New → Crv X Crv → Pick → Zoom in → *Select the involute profile → Select the pitch circle close to the placement of the point* **Refer Fig. 15.9.**
	15. Create a datum plane through the pitch point.	INSERT → DATUM → PLANE → Through → Pick → *Select the datum axis from the model tree* → Through → Pick → *Select PNT0 by clicking on the word "PNT0"* → Done VIEW → SAVED VIEWS → *Select Front* → SET → CLOSE **Refer Fig. 15.10.**

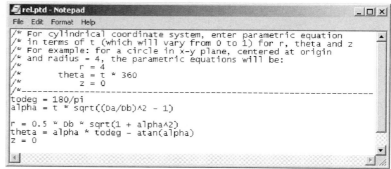

```
rel.ptd - Notepad
File  Edit  Format  Help
/* For cylindrical coordinate system, enter parametric equation
/* in terms of t (which will vary from 0 to 1) for r, theta and z
/* For example: for a circle in x-y plane, centered at origin
/* and radius = 4, the parametric equations will be:
/*         r = 4
/*     theta = t * 360
/*         z = 0
/*---------------------------------------------
todeg = 180/pi
alpha = t * sqrt((Da/Db)^2 - 1)

r = 0.5 * Db * sqrt(1 + alpha^2)
theta = alpha * todeg - atan(alpha)
z = 0
```

Fig. 15.7.

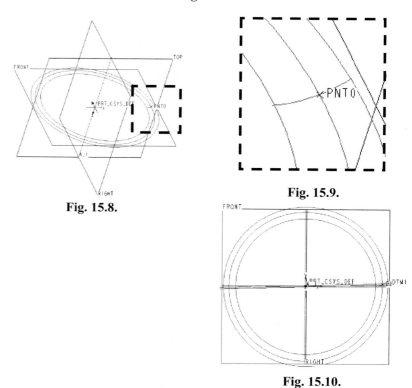

Fig. 15.9.

Fig. 15.8.

Fig. 15.10.

Goal	Step	Commands
Name the involute profile and the pitch point	16. Name the involute profile and the pitch point.	Set Up → Name → Feature → Pick → *Select the involute profile* → Involute_Profile → *Select the pitch point* → Pitch_Point → Done Sel → Done **Refer Fig. 15.11.**
Mirror the involute profile	17. Mirror the involute profile.	Feature → Copy → Mirror → Select → Dependent → Done → Select → Pick → *Select the involute profile from the model tree* → Done Sel → Done → Make Datum → Through → Pick → *Select the datum axis (A_1) from the model tree* → Angle → Plane → Pick → *Select DTM1 from the model tree* → Done → Enter Value → 90/N → ✓ **Refer Fig. 15.12.**
Create a profile for cutting the teeth	18. Start "Datum – Curve" feature.	INSERT → DATUM → CURVE → Sketch → Done
	19. Establish the sketching plane.	Use Prev → Okay

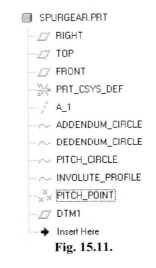

Fig. 15.11.

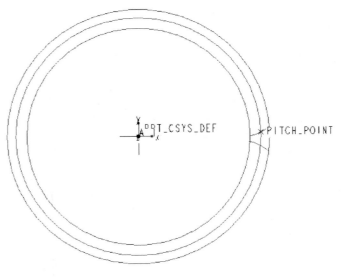

Fig. 15.12.

Goal	Step	Commands
Create a profile for cutting the teeth (Continued)	20. Establish new references.	**If references window is not visible, then open it using** SKETCH → REFERENCES **command.** *Select F2 (TOP) in the reference window →* Delete *→ Select F3(RIGHT) in the reference window →* Delete *→* ▸ *(in the "References" window) → Select the two involute profiles, and addendum and dedendum circles →* CLOSE **Refer Fig. 15.13.**
	21. Sketch the profile.	□ *→* Single *→ Select the two involute profiles →* Loop *→ Select the dedundum circle*
	22. Trim the additional lines from the dedendum circle.	EDIT → TRIM → DELETE SEGMENT (or ⌐) *→ Click on the top half of the circle away from the involute profile → Click on the bottom half of the circle away from the involute profile* **Refer Fig. 15.14.**
	23. Trim the additional lines from the involute profile.	Zoom in **(Refer Fig. 15.15)** *→ Remove the involute profile below the dedendum circle* **Refer Fig. 15.16.**
	24. Exit sketcher.	✔
	25. Accept the feature creation.	OK

Fig. 15.13.

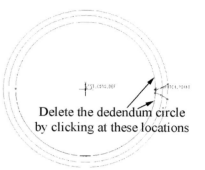

Delete the dedendum circle by clicking at these locations

Fig. 15.14.

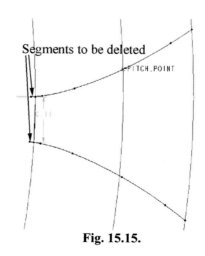

Segments to be deleted

Fig. 15.15.

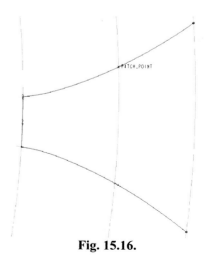

Fig. 15.16.

Goal	Step	Commands
Create a cutting surface	26. Start "Surface – Extrude" feature.	Create → Surface → New → Extrude → Done → One Side → Open Ends → Done
	27. Establish the sketching plane.	Use Prev → Okay
	28. Sketch the profile.	▢ → Loop → *Select the profile of the cut (the previous datum curve)*
	29. Exit sketcher.	✔
	30. Define the depth.	Blind → Done → F → ✔ → Yes → Yes
	31. Accept the feature creation.	Preview → VIEW → DEFAULT ORIENTATION → OK → Done/Return **Refer Fig. 15.17.**
Create a base cylinder	32. Start "Protrusion – Extrude" feature.	Create → Solid → Protrusion → Extrude → Solid → Done → One Side → Done
	33. Define the sketching plane.	Use Prev → Flip (The arrow should be along the surface) → Okay
	34. Create the outer circle.	▢ → Loop → *Select the addendum circle (outermost circle)*
	35. Exit sketcher.	✔
	36. Define the depth.	Blind → Done → F → ✔ → Yes → Yes
	37. Accept the feature creation.	Preview → VIEW → DEFAULT ORIENTATION → OK → ▣ → Done **Refer Fig. 15.18.**

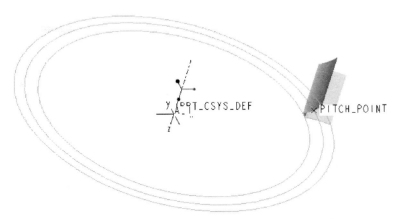

Fig. 15.17.

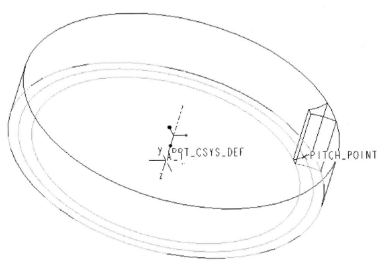

Fig. 15.18.

Goal	Step	Commands
Name the new features	38. Name the two new features.	▶ PART → Set Up → Name → Feature → Pick → *Select the cut profile* → Cut_Profile → *Select the cutting surface* → Cutting_Surface → *Select the protrusion* → Base_Cylinder → Done Sel → Done **Refer Fig. 15.19.**
Copy the cutting surface	39. Copy the cutting surface.	Feature → Create → Surface → Transform → Move → Copy → Done → Pick → *Query select the cutting surface (quilt) in the graphics window* → Done Sel → Rotate → Crv/Edg/Axis → Pick → *Select the datum axis* → Okay → 360/N → ✓ → Done Move → Done/Return **Refer Fig. 15.20.**
Create the cut	40. Create the initial cut.	Create → Solid → Cut → Use Quilt → Solid → Done **ProE opens "CUT: Use Quilt" window.** Pick → *Query select the **second** quilt surface from the graphics window* → Side 2 → ✓ **Refer Figs 15.21 and 15.22.**

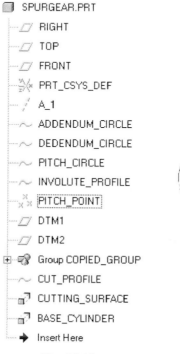

SPURGEAR.PRT
- RIGHT
- TOP
- FRONT
- PRT_CSYS_DEF
- A_1
- ADDENDUM_CIRCLE
- DEDENDUM_CIRCLE
- PITCH_CIRCLE
- INVOLUTE_PROFILE
- PITCH_POINT
- DTM1
- DTM2
- Group COPIED_GROUP
- CUT_PROFILE
- CUTTING_SURFACE
- BASE_CYLINDER
- ➔ Insert Here

Fig. 15.19.

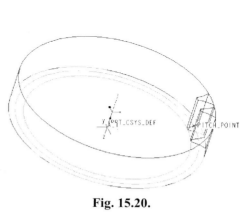

Fig. 15.20.

Fig. 15.21.

Fig. 15.22.

Goal	Step	Commands
Create the teeth	41. Group the cutting surface and the cut.	Group → Cancel → Local Group → Cut → ✓ → Select → Pick → *Select the last two features (Transformed surface and Cut) from the model tree* → Done Sel → Done
	42. Pattern the cut.	Pattern → Select → Pick → *Select the Group CUT from the model tree* → Value → *Click on the dimension 12.9* → (Enter the increment) 360/N → ✓ → Done → (Enter the number of instances) 28 → ✓ → Done → Done/Return **Refer Fig. 15.23.**
Save the file and exit ProE	43. Save the file and exit ProE.	FILE → SAVE → SPURGEAR.PRT → ✓ → FILE → EXIT → YES

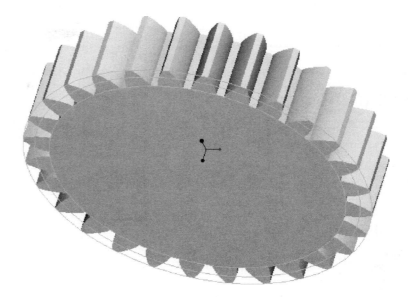

Fig. 15.23.

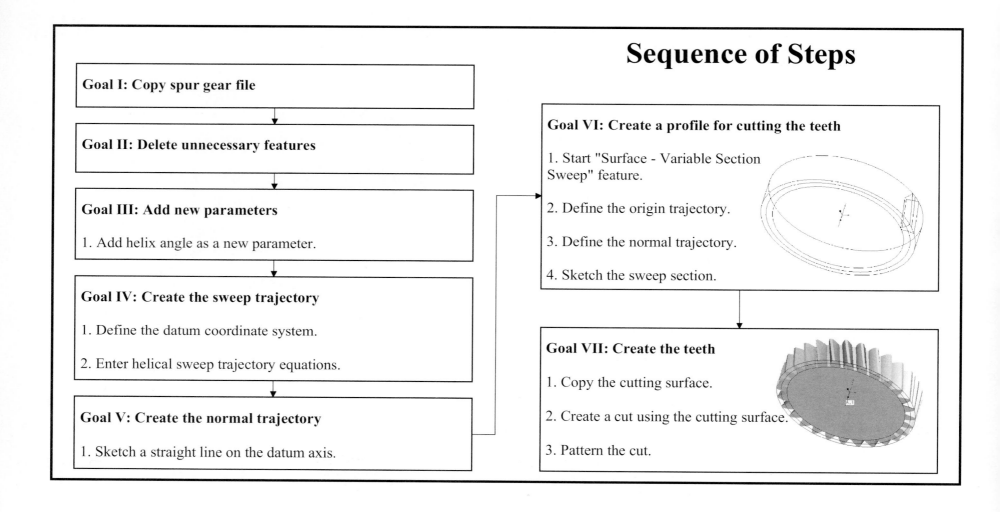

Sequence of Steps

Goal I: Copy spur gear file

Goal II: Delete unnecessary features

Goal III: Add new parameters

1. Add helix angle as a new parameter.

Goal IV: Create the sweep trajectory

1. Define the datum coordinate system.

2. Enter helical sweep trajectory equations.

Goal V: Create the normal trajectory

1. Sketch a straight line on the datum axis.

Goal VI: Create a profile for cutting the teeth

1. Start "Surface - Variable Section Sweep" feature.

2. Define the origin trajectory.

3. Define the normal trajectory.

4. Sketch the sweep section.

Goal VII: Create the teeth

1. Copy the cutting surface.

2. Create a cut using the cutting surface.

3. Pattern the cut.

Goal	Step	Commands
Open a new file for the helical gear part	1. Set up the working directory.	FILE → SET WORKING DIRECTORY → *Select the working directory* → OK
	2. Open spur gear part.	FILE → OPEN → spurgear → OPEN
	3. Save the part as the helical gear part.	FILE → SAVE A COPY → (New Name) HelicalGear → OK
	4. Open the helical gear part.	FILE → ERASE → CURRENT → YES → FILE → OPEN → *Select the HELICALGEAR.PRT* → OPEN
Delete unnecessary features	5. Reorder features.	*In the model tree, drag and drop the base cylinder before the cutting surface in the model tree window* **Refer Fig. 15.24.**
	6. Delete the cutting surface and the patterned cuts.	*Select the last two features (Refer Fig. 15.25)* → *Right Mouse* → *Delete* → OK **Refer Figs. 15.25 and 15.26.**

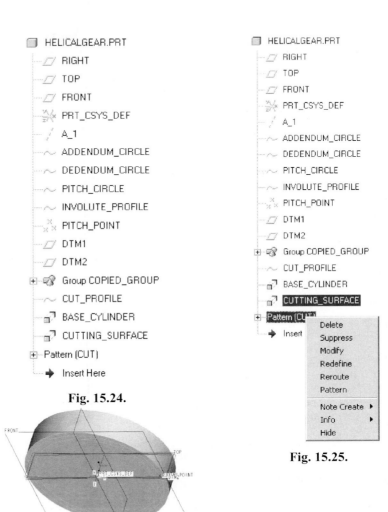

Fig. 15.24.

Fig. 15.25.

Fig. 15.26.

Goal	Step	Commands
Add new parameter	7. Add helix angle as a new parameter.	Set Up → Parameters → Part → Create → Real Number → beta (for helix angle) → ☑ → 20 → ☑ → Done
Create the sweep trajectory	8. Create the sweep trajectory datum curve.	INSERT → DATUM → CURVE → From Equation → Done → Select → Pick → *Select the default coordinate system* → Cylindrical
	9. Enter the equations for the helical sweep trajectory.	Type the equations in the equation editor. **Refer Fig. 15.27.** FILE → EXIT → YES
	10. Accept the feature creation after previewing.	Preview → VIEW → DEFAULT ORIENTATION → OK **Refer Fig. 15.28.**
Create a normal trajectory	11. Start "Datum – Curve" feature.	INSERT → DATUM → CURVE → Sketch → Done
	12. Define the sketcher plane.	Setup New → Plane → Pick → *Select the RIGHT datum plane* → Okay → TOP → *Select the TOP datum plane*
	13. Add new reference.	*Select the right face of the base cylinder*
	14. Sketch a straight line on the datum axis.	╲ → *Select points 1 and 2* → *Middle Mouse* **Refer Fig. 15.29.**
	15. Exit sketcher.	✔
	16. Accept the feature creation.	OK → VIEW → DEFAULT ORIENTATION

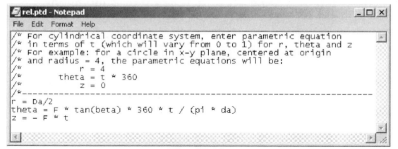

```
rel.ptd - Notepad
File  Edit  Format  Help
/* For cylindrical coordinate system, enter parametric equation
/* in terms of t (which will vary from 0 to 1) for r, theta and z
/* For example: for a circle in x-y plane, centered at origin
/* and radius = 4, the parametric equations will be:
/*          r = 4
/*          theta = t * 360
/*          z = 0
/*------------------------------------------------------------
r = Da/2
theta = F * tan(beta) * 360 * t / (pi * da)
z = - F * t
```

Fig. 15.27.

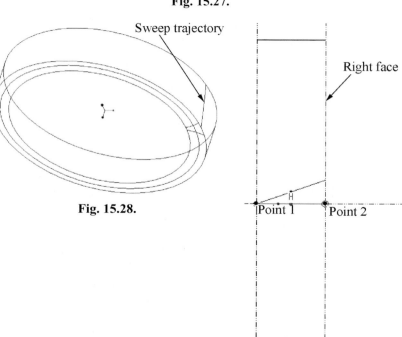

Sweep trajectory

Right face

Fig. 15.28.

Point 1   Point 2

Fig. 15.29.

Goal	Step	Commands
Create a cutting surface	17. Start "Surface – Variable Section Sweep" feature.	Feature → Create → Surface → New → Advanced → Done → Var Sec Swp → Done → Norm to Traj → Done
	18. Define the origin trajectory.	Select Traj → One By One → Select → Pick → *Select the helical sweep trajectory curve* → Done **Refer Fig. 15.30.**
	19. Define the normal trajectory.	Use Norm Traj → Done → Select Traj → One By One → Select → Pick → *Select the normal trajectory curve* → Done → Done
	20. Select the end type.	Open Ends → Done
	21. Sketch the sweep section using use edge.	□ → *Loop* → *Select the cut profile datum curve in the graphics window*
	22. Delete all references.	SKETCH → REFERENCES → *Select all the references in the reference window* → DELETE → CLOSE **Refer Fig. 15.31.** **New dimensions defining the curve should appear in the graphics window.**

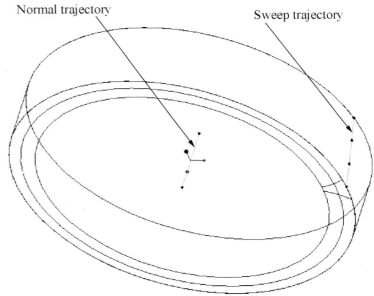

Normal trajectory · Sweep trajectory

Fig. 15.30.

Fig. 15.31

Goal	Step	Commands
Create a cutting surface (Continued)	23. Exit sketcher.	✔
	24. Accept the feature creation.	Preview → VIEW → DEFAULT ORIENTATION → OK → ⊞ → Done/Return **Refer Fig. 15.32.**
Copy the cutting surface	25. Copy the cutting surface.	Create → Surface → Transform → Move → Copy → Done → Pick → *Query select the cutting surface in the graphics window* → Done Sel → Rotate → Crv/Edg/Axis → Pick → Select the datum axis → Okay → 360/N → ✔ → Done Move → Done/Return **Refer Fig. 15.33.**
Create the teeth	26. Create the initial cut.	Create → Solid → Cut → Use Quilt → Solid → Done **ProE opens "CUT: Use Quilt" window.** Pick → *Query select the cutting surface from the graphics window* → ⟳ Side 2 → ✔ **Refer Fig. 15.34.**

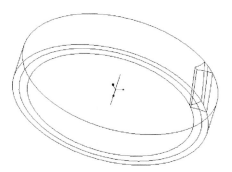

Fig. 15.32.

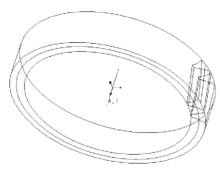

Fig. 15.33.

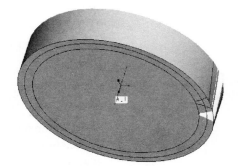

Fig. 15.34.

Goal	Step	Commands
Create the teeth (Continued)	27. Group the cutting surface and the cut.	Group → Cancel → Local Group → Cut → ✓ → Select → Pick → *Select the last two features (Transformed surface and Cut) from the model tree* → Done Sel → Done
	28. Pattern the cut.	Pattern → Select → Pick → *Select the Group CUT from the model tree* → Value → *Click on the dimension 12.9* → (Enter the increment) 360/N → ✓ → Done → (Enter the number of instances) 28 → ✓ → Done → Done/Return **Refer Fig. 15.35.**
Save the file and exit ProE	29. Save the file and exit ProE.	FILE → SAVE → HELICALGEAR.PRT → ✓ → FILE → EXIT → YES

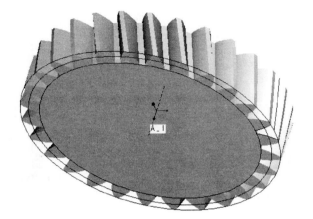

Fig. 15.35.

LESSON 16
SHAFT DRAWING

Learning Objectives:

- Learn to create **Drawing Format Sheets**.

- Learn to create **Orthographic** and **Trimetric Views** for a part.

- Learn **Dimensioning** and **Tolerancing** the part drawing.

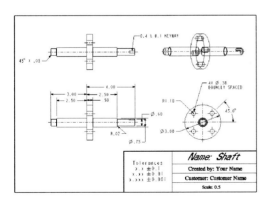

Design Information:

An engineering drawing has been the primary document exchanged between the design and manufacturing departments. It should accurately and unambiguously represent the part to avoid unexpected delays and disputes. The drawing should define the overall geometry, all the dimensions, and the required tolerances for producing the part. Manufacturing engineers follow these guidelines to manufacture the part. To facilitate easy manufacturing, it is important to add necessary views to accurately represent the part and increase clarity by properly arranging the dimensions and tolerances.

Sequence of Steps

Goal I: Create a drawing format

1. Create borders.

2. Create the title box.

3. Enter text in the title box.

4. Format the text in the title box.

5. Save the format sheet.

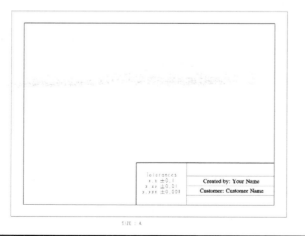

Goal II: Create a drawing for the casing part

1. Add the front view.

2. Add the top view.

3. Add the side view.

4. Create a trimetric view.

5. Move views.

6. Display centerlines.

7. Dimension all features.

8. Clean up the drawing by moving the dimensions.

9. Add name and scale to the title box.

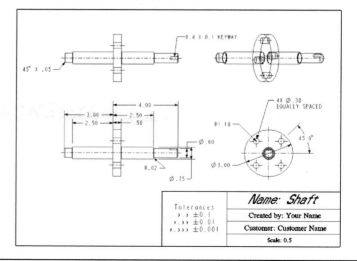

Goal	Step	Commands
Open a drawing file for the shaft part	1. Set up the working directory.	FILE → SET WORKING DIRECTORY → *Select the working directory* → OK
	2. Open a drawing file for the shaft part.	FILE → NEW → FORMAT → format_A→ OK → (Specify Templete) *Empty* → (Orientation) Landscape → (Standard Size) A → OK **Refer Fig. 16.1.**
Create borders	3. Create line 1.	SKETCH → CHAIN → SKETCH → LINE → SKETCH → SPECIFY ABS COORDS → (x) 0.5 → (Y) 0.5 → ✓ → SKETCH → SPECIFY ABS COORDS → (x) 10.5 → (y) 0.5 → ✓
	4. Create line 2.	(x) 10.5 → (y) 8 → ✓
	5. Create line 3.	(x) 0.5 → (y) 8 → ✓
	6. Create line 4.	(x) 0.5 → (y) 0.5 → ✓ → ✗ **Refer Fig. 16.2.**

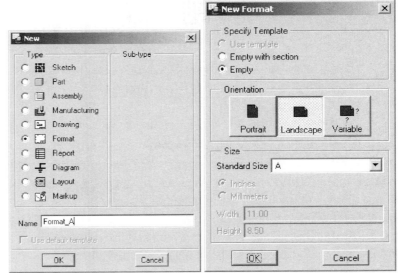

Fig. 16.1.

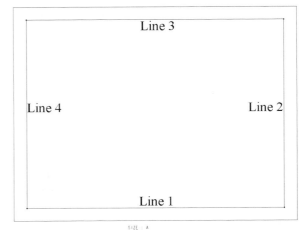

Fig. 16.2.

Goal	Step	Commands
Create the title box	7. Create line 5.	SPECIFY ABS COORDS → (x) <u>10.5</u> → (y) <u>2.25</u> → ✓ → SKETCH → SPECIFY ABS COORDS → (x) <u>5</u> → (y) <u>2.25</u> → ✓
	8. Create line 6.	(x) <u>5</u> → (y) <u>0.5</u> → ✓ → ☒
	9. Optional Step: If additional lines are created by mistake, then delete them.	↖ → *Select the items* → ***DELETE***
	10. Modify the line thickness.	FORMAT → LINE STYLE → Pick Many → Pick Box → Inside Box → *Draw a box to select the six lines* → Done Sel → (Width) 0.03 → Apply → Close → Done/Return **Refer Figs. 16.3 and 16.4.**
	11. Create the inside vertical line.	SKETCH → LINE → SKETCH → SPECIFY ABS COORDS → (x) <u>7</u> → (y) <u>0.5</u> → ✓ → SKETCH → SPECIFY ABS COORDS → (x) <u>7</u> → (y) <u>2.25</u> → ✓ → *Middle Mouse*
	12. Create the inside horizontal line 1.	SKETCH → SPECIFY ABS COORDS → (x) <u>7</u> → (y) <u>0.9</u> → ✓ → SKETCH → SPECIFY ABS COORDS → (x) <u>10.5</u> → (y) <u>0.9</u> → ✓ → *Middle Mouse*
	13. Create the inside horizontal line 2.	SKETCH → SPECIFY ABS COORDS → (x) <u>7</u> → (y) <u>1.3</u> → ✓ → SKETCH → SPECIFY ABS COORDS → (x) <u>10.5</u> → (y) <u>1.3</u> → ✓

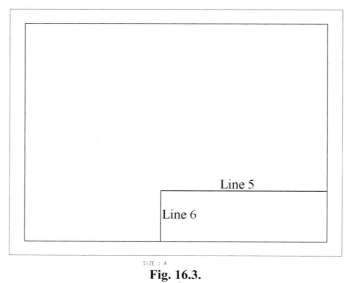

SIZE : A

Fig. 16.3.

Fig. 16.4.

Goal	Step	Commands
Create the title box (Continued)	14. Create the inside horizontal line 3.	SKETCH → LINE → SKETCH → SPECIFY ABS COORDS → (x) <u>7</u> → (y) <u>1.7</u> → ✔ → SKETCH → SPECIFY ABS COORDS → (x) <u>10.5</u> → (y) <u>1.7</u> → ✔ **Refer Fig. 16.5.**
Enter text in the title box	15. Create the text in the title box.	INSERT → NOTE → No Leader → Enter → Horizontal → Standard → Default → Make Note → Pick Pnt → *Select a point in the second column and second row* → <u>Created by: Your Name</u> → ✔ → ✔ → Make Note → Pick Pnt → *Select a point in the second column and third row* → <u>Customer: Customer Name</u> → ✔ → ✔ → Make Note → Pick Pnt → *Select a point in the first column* → <u>Tolerances</u> → ✔ → <u>x.x ±0.1</u> → ✔ → <u>x.xx ±0.01</u> → ✔ → <u>x.xxx ±0.001</u> → ✔ → ✔ → Done/Return **± symbol is available in the symbol palette.** **Refer Fig. 16.6.**

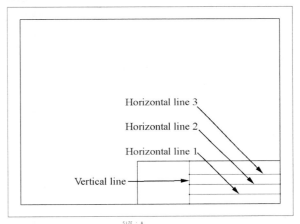

Fig. 16.5.

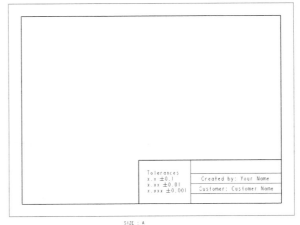

Fig. 16.6.

Goal	Step	Commands
Enter text in the title box (Continued)	16. Center justify the tolerance text.	FORMAT → TEXT STYLE → Pick → *Select the four tolerance text lines* → Done Sel → *(Justify Horiz) Center* → Apply → OK **Refer Fig. 16.7.**
	17. Move the text to the desired location.	▸ → Pick → *Select the text, move it to the desired location and then, click again to place the text* **Refer Fig. 16.8.**
Change the font	18. Change the font.	FORMAT → TEXT STYLE → *Select the text lines* → Done Sel → *(Font) CG Times* → Apply → OK **Refer Fig. 16.8.**
Save drawing format & erase the current session	19. Save the drawing format.	FILE → SAVE → FormatA.FRM → ✓
	20. Erase the current session.	FILE → ERASE → CURRENT → YES

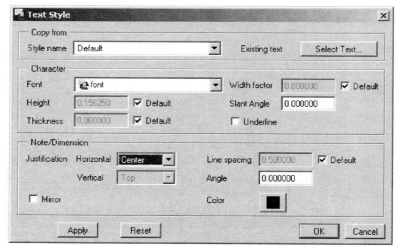

Fig. 16.7.

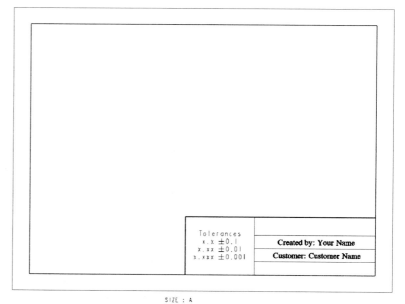

Fig. 16.8.

Goal	Step	Commands
Open a new file for the drawing	21. Open a new drawing file with the format created earlier.	FILE → NEW → DRAWING → <u>SHAFT</u> → OK → *(Default Model) Select the shaft.prt file → Empty with format → FormatA.frm →* OK **Refer Fig. 16.9.**
Create the front view	22. Start creating the front view.	Views → Add View → General → Full View → No Xsec → Scale → Done → *Click in the lower left quadrant of the drawing →* <u>0.5</u> → ✓
	23. Orient the front view.	*(Front) Select the FRONT datum plane →* (Top) *Select the TOP datum plane →* OK **Refer Fig. 16.11.**
Create the top view	24. Create the top view.	Add View → Projection → Full View → No Xsec → No Scale → Done → *Click in the top left quadrant (Top view position)*
Create the side view	25. Create the side view.	Add View → Projection → Full View → No Xsec → No Scale → Done → *Click in the bottom right quadrant (Side view position)* **Refer Fig. 16.12.**

Fig. 16.9.

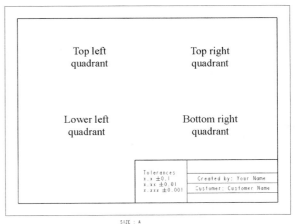

Fig. 16.10.

Fig. 16.11.

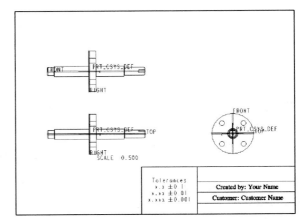

Fig. 16.12.

Goal	Step	Commands
Create a trimetric view	26. Create a trimetric view for visualization.	Add View → General → Full View → No Xsec → Scale → Done → *Click in the top right quadrant of the drawing sheet* → 0.5 → ✓ → (Type) Preference → (Default Orientation) Trimetric → OK → Done/Return **Refer Figs. 16.13 and 16.14.**
Move views	27. Move the views to create an appropriate layout.	Move View → *Select and move each view* → Done/Return
Display centerlines	28. Show all centerlines.	VIEW → SHOW AND ERASE → SHOW → ----A.1 → Show All → Yes → Done Sel
Dimension all features	29. Show all dimensions.	SHOW → ⊢1.2⊣ → SHOW ALL → YES → ACCEPT ALL → CLOSE → Done Sel **Reader can avoid the cluttering of dimensions by picking one feature at a time, and then showing its dimensions.**
Clean up the drawing	30. Turn off datum planes.	▱ ╱ ✱ ▱ *(Turn off datum planes and coordinate system)* → VIEW → REPAINT
	31. Change the display type to the hidden view.	⊞ **Refer Fig. 16.15.**

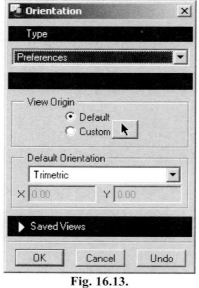

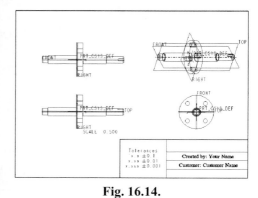

Fig. 16.14.

Fig. 16.13.

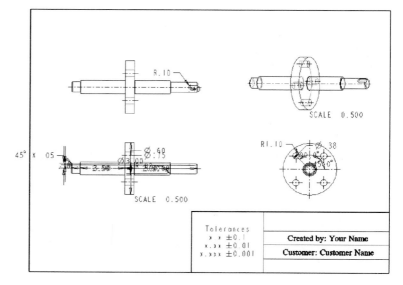

Fig. 16.15.

Goal	Step	Commands
Clean up the drawing (Continued)	32. Set drawing parameters.	Advanced → Draw Setup → (Option) draw_arrow_length → (Value) 0.1 → ADD/CHANGE → (Option) draw_arrow_style → (Value) Filled → ADD/CHANGE → drawing_text_height → (Value) 0.125 → ADD/CHANGE → APPLY → CLOSE → Done/Return **Refer Fig. 16.16.** **These preferences can be saved as a drawing set up file (.DTL).**
	33. Move the dimensions so that they do not overlap with each other.	↖ → Pick → *Select and move each dimension*
	34. Flip arrows if the space between them is too narrow.	↖ → *Select the dimension → Right Mouse → Flip Arrow*
	35. Switch dimensions across views. Try to place dimensions between the views as far as possible.	↖ → *Select the dimension → Right Mouse → Switch View → Click on the view to which the dimensions should be switched*
	36. Provide a gap between the witness lines and the geometry.	↖ → *Select the dimension → Select the witness line end and move it*

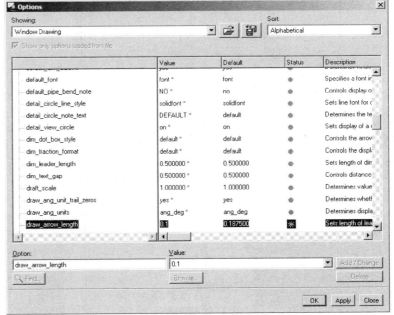

Fig. 16.16.

Goal	Step	Commands
Clean up the drawing (Continued)	37. Erase the unnecessary dimensions.	VIEW → SHOW AND ERASE → ERASE → ⊢1.2→ → *Select the 90 degree angle dimension, and the dimensions of the keyway* → Done Sel → CLOSE
	38. Modify the bolthole text.	↖ → *Select the dimension* → *Right Mouse* → *Properties* → Dimension Text Tab → <u>Enter text shown in Fig. 16.17.</u> → OK
	39. Make note describing the keyway.	INSERT → NOTE → Leader → Enter → Horizontal → Standard → Default → Make Note → Pick → *Select the keyway edge* → Done Sel → Done → Pick → *Select location for the note* → <u>0.4 X 0.1 KEYWAY</u> → ✓ → ✓

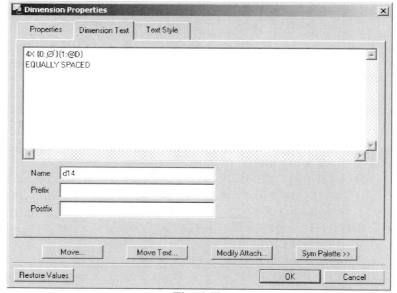

Fig. 6.17.

Goal	Step	Commands
Add part name and the scale	40. Add part name and scale in the title box.	INSERT → NOTE → No Leader → Enter → Horizontal → Standard → Default → Make Note → Pick Pnt → *Select location for the note* → Name: Shaft → ✓ → ✓ → Make Note → Pick Pnt → *Select location for the note* → Scale: 0.5 → ✓ → ✓
	41. Change the font.	FORMAT → TEXT STYLE → Pick → *Click on the text "Shaft"* → Done Sel → (Font) *Filled* → *Deselect two defaults next to height and thickness* → (Height) 0.3 → (Width) 1.2 → (Thickness) 0.02 → (Slant Angle) 30 → OK → Done/Return

Refer Fig. 16.18. |

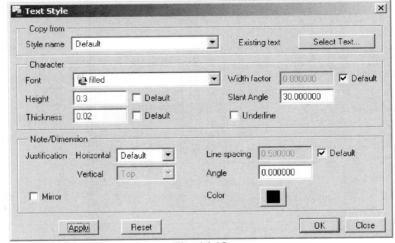

Fig. 16.18.

Goal	Step	Commands
Open the shaft part	42. Open the shaft part file.	FILE → OPEN → *Select shaft.prt file* → OPEN
Modify dimensions	43. Modify the dimensions.	Feature → Redefine → Select → Pick → *Select the first protrusion feature from the model tree* → *Select section from element list* → Define → **Refer Fig. 6.19.** Sketch → ⬆ → *Modify the two length dimensions as shown in Fig. 6.19.* → ✓ → OK → ▶ PART → Regenerate **Refer Fig. 6.20.**
	44. Save the changes.	FILE → SAVE → SHAFT.PRT → ✅
Switch to the drawing window	45. Switch to the drawing window.	WINDOW → SHAFT.DRW The changes made in the part mode can be seen in the drawing mode. **Refer Fig. 16.21.**
Save the file and exit ProE	46. Save the file and exit ProE.	FILE → SAVE → SHAFT.DRW → ✅ → FILE → EXIT → Yes

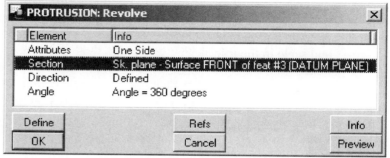

Fig. 16.19.

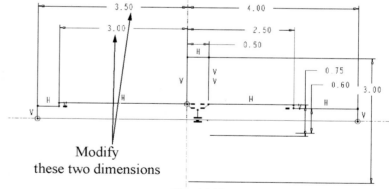

Modify
these two dimensions

Fig. 16.20.

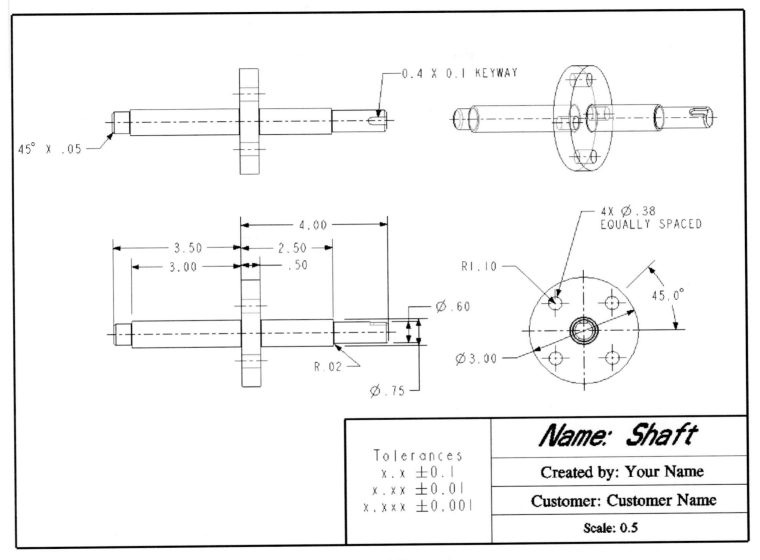

0.4 X 0.1 KEYWAY

45° X .05

4X Ø.38
EQUALLY SPACED

R1.10

45.0°

Ø.60

R.02

Ø3.00

Ø.75

4.00

3.50

2.50

3.00

.50

Tolerances
x.x ±0.1
x.xx ±0.01
x.xxx ±0.001

Name: Shaft

Created by: Your Name

Customer: Customer Name

Scale: 0.5

Fig. 16.21.

Lesson 17
Housing

Learning Objectives:

- Practice the key commands used in the part mode.

- Create *General*, *Sectional* and *Detailed* views.

- Control the display of the views.

- Practice *Show/Erase* command.

Design Information:

In this lesson, we will create a housing. We will then attempt to create an engineering drawing. To this end, we will start with defining the front and top views. Then, we will search for a proper sectional side view that shows maximum details in one view. During this quest, we will explore and learn full-, half- and offset-sectional views. We will add a detailed view to enlarge small details. The lesson ends with a detailed discussion on how to create other types of views.

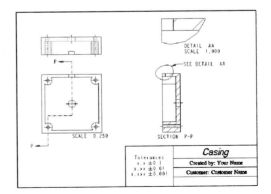

Sequence of Steps

Goal I: Create the housing part

1. Create the base feature.

2. Create the bolt holes.

3. Shell the inside surface.

4. Create a cut.

5. Create a central hole.

6. Create an offset section.

Goal II: Create a drawing for the housing part

1. Add the front view.

2. Add the top view.

3. Add the full sectional view.

4. Add the half-sectional view.

5. Add the offset sectional view.

6. Add the detailed view.

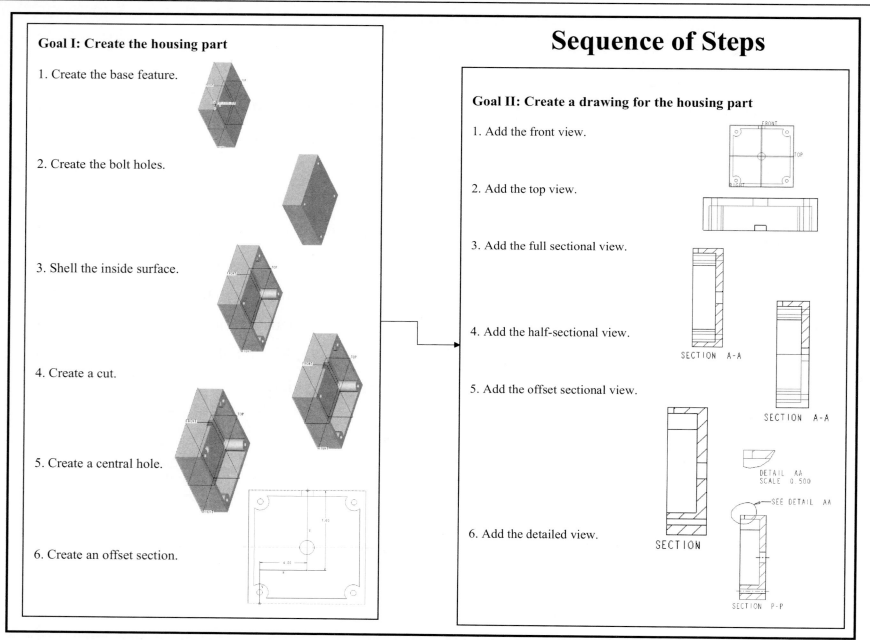

Goal	Step	Commands
Open a new file for the housing part	1. Set up the working directory.	FILE → SET WORKING DIRECTORY → *Select the working directory* → OK
	2. Open a new file for the housing part	FILE → NEW → *Part* → *Solid* → housing → OK
Create the base feature	3. Start "Protrusion – Extrude" feature.	Feature → Create → Solid → Protrusion → Extrude → Solid → Done
	4. Define the direction of extrusion.	One side → Done
	5. Select the sketching plane.	Setup New → Plane → Pick → *Select the RIGHT datum plane* → Okay
	6. Orient the sketching plane.	Top → Plane → Pick → *Select the TOP datum plane*
	7. Sketch the section.	▢ → *Select points 1 and 2* → *Middle mouse* **Refer Fig. 17.1.**
	8. Modify the dimensions.	▶ → *Double click each dimension and enter corresponding value* **Refer Fig. 17.1.**
	9. Exit sketcher.	✔
	10. Define the depth.	Blind → Done → 3 → ✔
	11. Accept the feature creation after previewing.	Preview → VIEW → DEFAULT ORIENTATION → OK **Refer Fig. 17.2.**

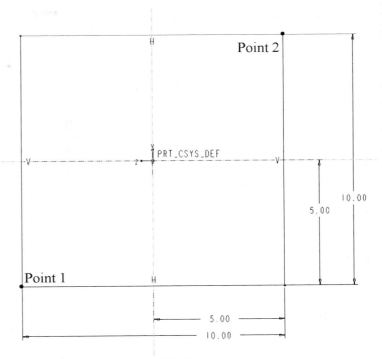

Fig. 17.1.

Fig. 17.2.

Goal	Step	Commands
Create the bolt holes	12. Create a corner hole.	Create → Solid → Hole → (Diameter) 0.5 → (Depth One) Thru All → (Primary Reference) *Select the plane parallel to the RIGHT datum plane* → (Linear Reference) *Select the top surface of the base feature* → (Distance) 1 → (Linear Reference) *Select the front surface of the base feature* → (Distance) 1 → ✓ **Refer Figs. 17.3 and 17.4.**
	13. Pattern the hole.	Pattern → Select → Pick → *Select the hole from the model tree* → Identical → Done → Value → *Select the horizontal dimension (1.0)* → 8 → ✓ → Done → 2 → ✓ → Value → *Select the vertical dimension (1.0)* → 8 → ✓ → Done → 2 → ✓ **Refer Fig. 17.5.**

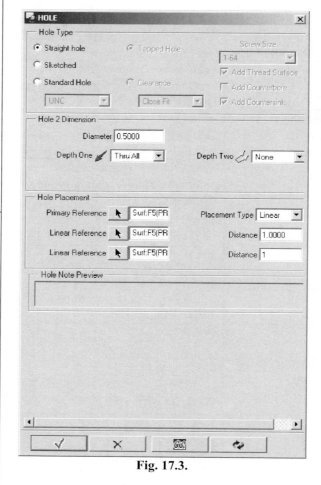

Fig. 17.3.

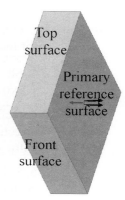

Top surface

Primary reference surface

Front surface

Fig. 17.4.

Fig. 17.5.

Goal	Step	Commands
Remove the inside material	14. Apply shell command.	Create → Solid → Shell → Pick → *Select surface 1 (Refer Fig. 17.6.)* → Done Sel → Done Refs → 0.5 → ✓ → Select *Spec Thick in the SHELL window* → DEFINE → Set Thickness → Pick → *Query select surface 2 (Surface that is lying on the RIGHT datum plane)* → 0.75 → ✓ → Done Sel → Done → OK **Refer Fig. 17.7.**
Create a cut	15. Start "Cut – Extrude" feature.	Create → Solid → Cut → Extrude → Solid → Done
	16. Define the cut direction.	One Side → Done
	17. Select the sketching plane.	Setup New → Plane → Pick → *Query select surface 3* → Okay **Refer Fig. 17.8.**
	18. Orient the sketching plane.	Right → Plane → Pick → *Select the RIGHT datum plane*
	19. Add a new reference.	*Select the right edge of the protrusion*
	20. Sketch the section.	▱ → *Select points 1 and 2* → *Middle Mouse* **Refer Fig. 17.8.**
	21. Modify the dimensions.	↖ → *Double click each the dimension and enter the corresponding value* **Refer Fig. 17.9.**

Fig. 17.6.

Fig. 17.7.

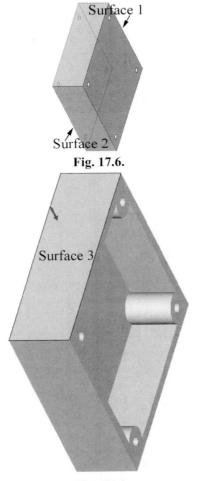

Fig. 17.8.

Fig. 17.9.

Goal	Step	Commands
Create a cut (Continued)	22. Exit sketcher.	✔
	23. Define the direction of material removal.	Okay
	24. Define the depth.	Through Next → Done
	25. Accept the feature creation.	Preview → OK **Refer Fig. 17.10.**
Create a central hole	26. Create a central hole.	Create → Solid → Hole → (Diameter) 1.25 → (Depth One) Thru All → (Primary Reference) *Select surface 1 (Refer Fig. 17.11)* → (Linear Reference) *Select the TOP datum plane* → (Distance) <u>0</u> → (Linear Reference) *Select the FRONT datum plane* → (Distance) <u>0</u> → ✔ → Done **Refer Fig. 17.12.**
Create an offset section	27. Create an offset section.	X-section → Create → Offset → Both Sides → Single → Done → <u>P</u> → ✔

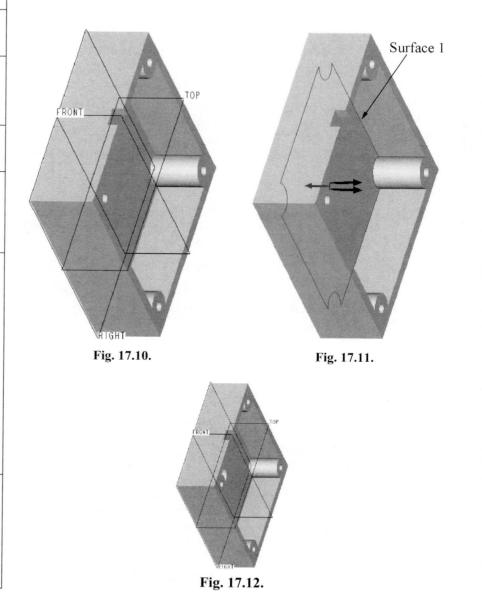

Fig. 17.10.   **Fig. 17.11.**

Surface 1

Fig. 17.12.

Goal	Step	Commands
Create an offset section (Continued)	28. Set up and orient the sketching plane.	Setup New → Plane → Pick → *Select the RIGHT datum plane →* Okay → Top → Plane → Pick → *Select the TOP datum plane*
	29. Sketch the cutting plane.	＼ → Select points 1, 2, 3, and 4 → *Middle Mouse* → (Align point 1 to the top line) YES → (Align point 4 to the bottom line) YES **Refer Fig. 17.13.**
	30. Modify the dimensions.	↖ → *Double click each dimension and enter the corresponding value* **Refer Fig. 17.13.**
	31. Exit sketcher.	✔ → Done/Return
Save the part	32. Save the part.	FILE → SAVE → HOUSING.PRT → ✔
Open a drawing file for the housing part	33. Open a new drawing file with the format created in the previous chapter.	FILE → NEW → DRAWING → HOUSING → OK → (Default Model) *housing.prt* → *Empty with format* → Format_A.frm → OK
Create the front view	34. Start creating the front view.	Views → Add View → General → Full View → No Xsec → Scale → Done → *Click in the lower left quadrant of the drawing* → 0.25 → ✔
	35. Orient the front view.	(Front) *Select the RIGHT datum plane* → (Top) *Select the TOP datum plane* → OK **Refer Figs. 17.14 and 17.15.**

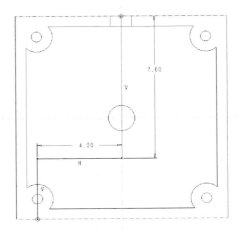

Fig. 17.13.

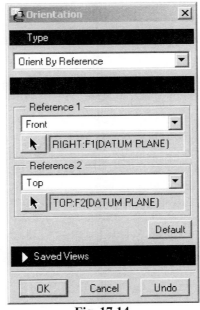

Fig. 17.14.

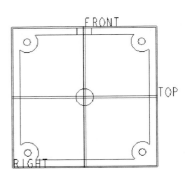

Fig. 17.15.

Goal	Step	Commands
Create the top view	36. Create the top view.	Add View → Projection → Full View → No Xsec → No Scale → Done → *Click in the top left quadrant* (Top view position) **Refer Fig. 17.16.**
Create the right section view	37. Create the right section view.	Add View → Projection → Full View → Section → No Scale → Done → Full → Total Xsec → Done → *Click in the bottom right quadrant* (Right view position) → Create → Planar → Single → Done → A → ✓ → Plane → Pick → *Select the FRONT datum plane in the front view* → *Select the front view* **Refer Fig. 17.17.**
	38. Delete this view, as it does not provide any additional information.	Delete View → *Select the sectional view* → Confirm → Done Sel
	39. Add a half-section view.	Add View → Projection → Full View → Section → No Scale → Done → Half → Total Xsec → Done → *Click in the bottom right quadrant* (Right view position) → → Pick → *Select the TOP datum plane* **(The section appears on one side of this plane)** → Okay → Retrieve → A → *Select the front view* **Refer Fig. 17.18.**

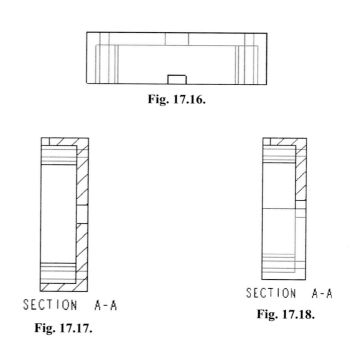

Fig. 17.16.

SECTION A-A

Fig. 17.17.

SECTION A-A

Fig. 17.18.

ProE associates the sections with the part file. Therefore, the sections can be created, deleted, viewed, modified or erased in the part mode. "X-section" command in the part mode provides the access to the sections.

Goal	Step	Commands
Create the right section view (Continued)	40. Delete the view as an offset view can provide more information.	Delete View → *Select the sectional view* → Confirm → Done Sel
	41. Turn off the datum planes, axes, points and coordinate system.	Click on the following icons
	42. Create an offset section.	Add View → Projection → Full View → Section → No Scale → Done → Full → Total Xsec → Done → *Click in the right bottom quadrant* (Right view position) → → Retrieve → P → *Select the front view* → Okay **Refer Fig. 17.19.**
	43. Set display mode.	Disp Mode → View Disp → Pick → *Select the three views* → Done Sel → Hidden Line → Tan Default → Hide Skeleton → Done **Refer Fig. 17.20.**
	44. Turn off the hidden lines in the sectional view.	Edge Disp → Erase Line → Tan Default → Any View → Pick → Zoom in → *Select the six hidden lines (seen in gray color)* → Done Sel → Done → Done/Return **Refer Fig. 17.21.**

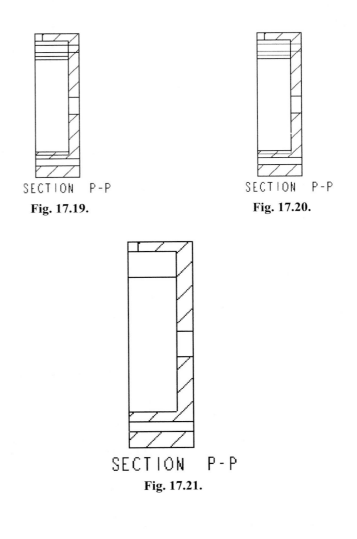

SECTION P-P

Fig. 17.19.

SECTION P-P

Fig. 17.20.

SECTION   P-P

Fig. 17.21.

Goal	Step	Commands
Create the right section view (Continued)	45. Show centerlines.	VIEW → SHOW AND ERASE → Show → ----A_1 → Show All → Yes → Done Sel → Sel to Remove → *Select unnecessary axes* → Done Sel → CLOSE **Refer Fig. 17.22.**
Create a detailed view	46. Create a detailed view to increase the visibility of the notch.	Views → Add View → Detailed → Full View → No Sec → Scale → Done → *Select the top right quadrant* → 0.5 → ✓ → *Select point 1 (Refer Fig. 17.23.)* → *Sketch a spline (Middle Mouse to discontinue the spline creation)* → (Name) AA → ✓ → Spline → *Select a point on the spline* → Pick Pnt → *Select a point to place the note* **Refer Fig. 17.24.**
Modify scale	47. Modify the scale.	Modify View → Change Scale → Pick → *Select the detailed view* → 1 → ✓ → Done/Return
Save the file and exit ProE	48. Save the file and exit ProE.	FILE → SAVE → HOUSING.DRW → ✓ → FILE → EXIT → Yes

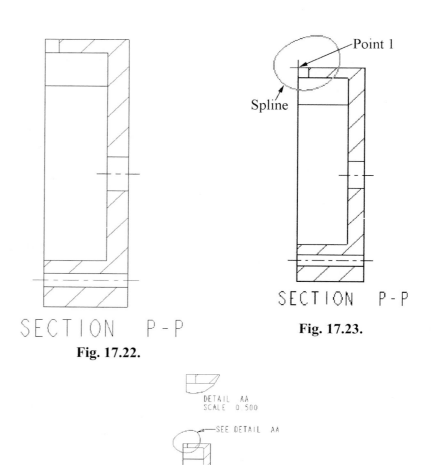

SECTION P-P

Fig. 17.22.

Point 1

Spline

SECTION P-P

Fig. 17.23.

DETAIL AA
SCALE 0.500

SEE DETAIL AA

SECTION P-P

Fig. 17.24.

Auxiliary Views

Auxiliary views are used to show the true geometry and dimensions of a geometric entity that is not parallel to any of the principle planes (Front, Top and Side). This is primarily used to show the true geometry of inclined surfaces. Auxiliary views can be created by using the following commands:

Views → Add View → Auxiliary →
Determines the view type as an auxiliary view.

Full View/Half-View/Partial → No Xsec/Section → No Scale → Done
Defines the view parameters.

Select the center point for the auxiliary view → Pick → *Select an edge or an axis through, edge of or datum plane from which to project the view*

(For partial views) Sketch a closed spline on the partial view

This set of commands creates an auxiliary view.

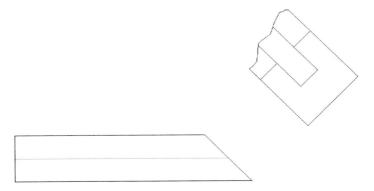

Fig. 17.25. (Partial auxiliary view of a C-channel).

Broken View

When the aspect ratio (overall length to section width) of a part is very large, the broken view is the most effective drawing technique. For instance, only parts of the long channel can be shown without sacrificing any design details. Broken views can be created by using the following commands:

Views → Add View → General → Broken View → No Section → No Scale → Done
Determines the view type as a broken view and establishes the view parameters.

Select the center point for the view → *Orient the view by picking the front and top surfaces* → OK →

Add → Vertical → Pick

Select two points defining the break lines → Done
Note that the two parts of the bar/beam will be together at this point.

Select the vertical/horizontal break lines → Done Sel → Heart Beat → View Outline → Done Sel → Done

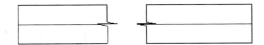

Fig. 17.26.

Revolved Section

Revolved sections are used to show the cross section of bar or beam. It can be constructed by using the following commands:

Views → Add View → Revolved
Determines the view type as a revolved view.

Full View → Section → No Scale → Done
Determines the view parameters.

Select the location for the revolved view → *Select the parent view for the revolved view* → Create/Retrieve a section →*Select the axis or plane of symmetry*

Fig. 17.27.

LESSON 18
CAM FOLLOWER ASSEMBLY

Learning Objectives:

- Explore the basics of assembly.

- Learn advanced features such as *Repeat* and *Pattern*.

- *Suppress* and *Resume* components in assembly.

- *Animate* components in assembly.

- Create parts in the assembly mode.

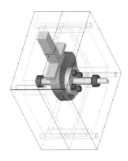

Design Information:

The components created in the previous lessons fit together to form the cam follower assembly. In this lesson, we will first create the cam subassembly, which uses the shaft, bearings, bolts and nuts. This subassembly is then assembled into the cam follower assembly. By creating appropriate relationships, we can dynamically rotate the cam subassembly. The housings and follower are assembled. The follower constraints ensure that it translates up and down when the cam assembly rotates. These simple motion checks ensure that there is no motion interference in the final assembly.

Sequence of Steps

Goal I: Create the cam assembly

1. Assemble the shaft.

2. Assemble the bearings.

3. Assemble the cam.

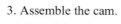

4. Assemble the bolts.

5. Assemble the nuts.

6. Create an exploded view.

7. Set col

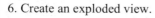

Goal II: Create the cam follower assembly

1. Assemble the cam subassembly at an angle.

2. Add relationships to rotate the cam subassembly.

3. Suppress the cam subassembly.

4. Assemble the left housing.

5. Assemble the right housing.

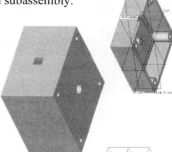

6. Resume all components.

7. Set up model display.

8. Suppress the housing parts.

9. Create a datum point at the apex of the cam.

10. Create the follower part in the assembly.

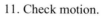

11. Check motion.

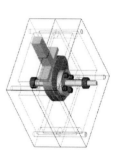

Goal	Step	Commands
Open a new file for the cam subassembly	1. Set up the working directory.	FILE → SET WORKING DIRECTORY → *Select the working directory* → OK
	2. Open a new file for the cam assembly.	FILE → NEW → Assembly → Design → Cam → OK
Assemble the shaft	3. Start assembling the shaft part.	Component → Assemble → *Select shaft.prt* → Open
	4. Establish constraints.	*Select Default under constraint type by clicking on its icon* **Refer Fig. 18.1.** **The placement status message in the component placement window should show "fully constrained."**
	5. Place the shaft part.	OK **Refer Fig. 18.2.**

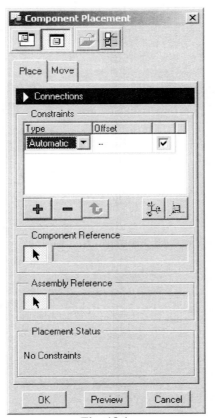

Fig. 18.1.

Constraint types

In ProE, constraints are used to assemble a part. Some common constraints are:

ALIGN: Lines up the selected surfaces or axes. In other words, the constraint makes the surfaces coplanar and the axes collinear.

MATE: Makes the surfaces coincident facing each other. Variations of these constraints include Mate Offset and Align Offset wherein an offset is introduced between the surfaces.

DEFAULT: Automatically aligns the default part coordinate system with that of the assembly.

INSERT: Makes the axes of two revolved surfaces (example: bolt and bolt hole) coincident.

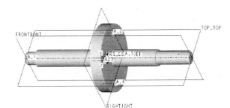

Fig. 18.2.

Goal	Step	Commands
Assemble the left bearing	6. Start assembling the bearing part.	Assemble → *Select bearing.prt* → Open
	7. Turn off datum planes and coordinate systems.	
	8. Align the axes.	(Constraint type) Align → (Component reference) *Query select the bearing axis* → (Assembly reference) *Query select the shaft axis* **Refer Figs. 18.3 and 18.4.**
	9. Mate the bearing surface with the shoulder.	(Constraint type) Mate → (Component reference) *Query select the right flat face of the bearing* → (Assembly reference) *Query select the left bearing shoulder* → 0 → ✓ **Refer Figs. 18.3 and 18.4.**
	10. Place the shaft part.	OK **Refer Fig. 18.5.**

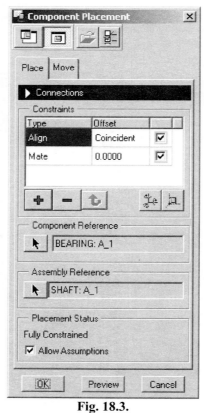

Fig. 18.3.

Display component in option

SEPARATE WINDOW: Opens the component in a separate window. The constraint type automatic is not available in this option.
ASSEMBLY: Displays the component is the assembly window. The component placement is updated with the addition of each constraint.

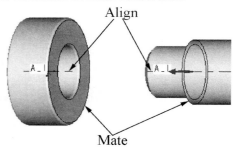

Fig. 18.4.

Fig. 18.5.

Goal	Step	Commands
Assemble the right bearing	11. Repeat the previous bearing.	Adv Utils → Repeat → *Select the bearing part from the model tree* → *Select the mate constraint as it has to be changed in the "Variable Assemble Refs" box* → ADD → Select → Pick → *Query Select the right bearing shoulder* → CONFIRM → Done/Return **Refer Figs. 18.6, 18.7 and 18.8.**
Assemble the plate cam	12. Start assembling the cam part.	Assemble → *Select platecam.prt* → Open
	13. Align the axes.	(Constraint type) Align → (Component reference) *Select the cam central hole axis* → (Assembly reference) *Select the shaft axis*
	14. Mate the cam surface with its shoulder.	(Constraint type) Mate → (Component reference) *Query select the cam surface where the hole is ending* → (Assembly reference) *Query select the left surface of the shoulder* → 0 → ✓ **Refer Fig. 18.9.**
	15. Place the shaft part.	OK **Refer Fig. 18.10.**

Fig. 18.6.

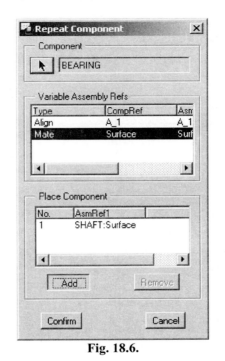

Fig. 18.9.

Repeat

Repeat allows the user to assemble a component at multiple locations. This command simplifies the assembly process by requiring the user to define only the new constraints.

Right bearing shoulder
Fig. 18.7.

Fig. 18.8.

Fig. 18.10.

Goal	Step	Commands
Assemble the bolts	16. Start assembling the bolt part.	Assemble → *Select bolt.prt* → Open
	17. Insert the bolt.	(Constraint type) Insert → (Component reference) *Query select the shank surface* → (Assembly reference) *Query select the surface of the bolt hole in the shaft part*
	18. Mate the bottom surface of the bolt head with the bottom surface of the countersunk hole.	(Constraint type) Mate → (Component reference) *Query select the bottom surface of the bolt head* → (Assembly reference) *Query select the bottom surface of the countersunk hole* → 0 → ✓ **Refer Figs. 18.11 and 18.12.**
	19. Place the bolt part.	OK **Refer Fig. 18.13.**
	20. Pattern the bolt.	Pattern → Select → Pick → *Select the bolt part from the model tree* → Ref Pattern → Done **Refer Fig. 18.14.**
Assemble the nuts	21. Start assembling the nut part.	Assemble → *Select nut.prt* → Open
	22. Align the axis of the nut with that of the bolt.	(Constraint type) Align → (Component reference) *Query select the axis of the nut* → (Assembly reference) *Query select the axis of the bolt*

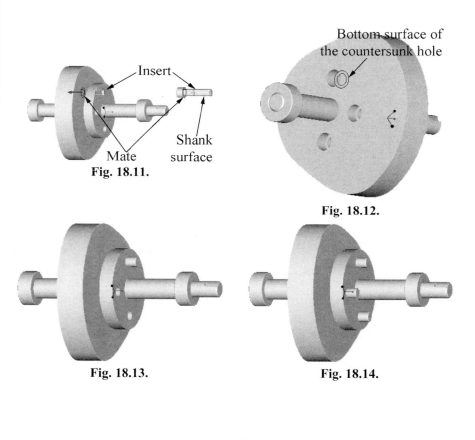

Insert

Bottom surface of the countersunk hole

Mate   Shank surface

Fig. 18.11.

Fig. 18.12.

Fig. 18.13.

Fig. 18.14.

Goal	Step	Commands
Assemble the nuts (Continued)	23. Mate the left surface of the nut with the right surface of the shaft shoulder.	(Constraint type) Mate → (Component reference) *Select a nut face* → (Assembly reference) *Select the right surface of the shoulder* → 0 → ✓
	24. Place the nut part.	OK **Refer Fig. 18.15.**
	25. Pattern the nut.	Pattern → Select → Pick → *Select the nut part from the model tree* → Ref Pattern → Done → Done/Return **Refer Fig. 18.16.**
Create an exploded view	26. Check the current exploded view.	VIEW → EXPLODE **Refer Fig. 18.17.** VIEW → UNEXPLODE
	27. Start creating a custom exploded view.	Explode State → Create → EXP0001 → ✓
	28. Select the axis for the reference motion.	*Select the shaft axis*
	29. Select each part and translate it along the axis.	*Select each part and move it along the axis* → OK → Done/Return → Done/Return **Refer Fig. 18.18.**

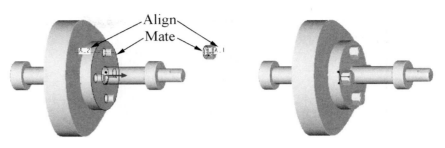

Fig. 18.15.   Fig. 18.16.

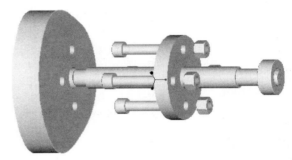

Fig. 18.17.

Fig. 18.18.

Goal	Step	Commands
Set up color	30. Define the model color.	VIEW → MODEL SETUP → COLOR AND APPEARANCES → ADD → *Click on the white area* → **Refer Fig. 18.19.** **ProE opens the "Color Editor" window.** **Refer Fig. 18.20.** *Click on Color wheel* → **ProE opens the color wheel.** *Select a suitable color* → CLOSE → ADD → *Add five colors* → OK → *Select a color* → (In the set object appearance box – Drop-down menu) Component → **Refer Fig. 18.21.** Pick → *Select the components* → Done Sel → Set → Set different colors to the components → Close
Save the subassembly file	31. Save the file.	FILE → SAVE → CAM.ASM → ✓

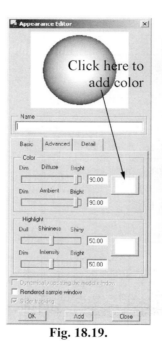

Fig. 18.19.

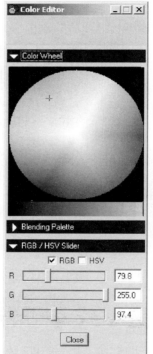

Fig. 18.20.

Fig. 18.21.

Goal	Step	Commands
Open a new file for the cam follower assembly	32. Open a new file for the cam follower assembly.	FILE → NEW → Assembly → Design → CamFollower → OK
Create a datum axis	33. Create a datum axis.	INSERT → DATUM → AXIS → Two Planes → Plane → Pick → *Select ASM_TOP and ASM_FRONT datum planes* **Refer Fig. 18.22.**
Create a datum plane	34. Create a datum plane.	INSERT → DATUM → PLANE → Through → Axis Edge Curve → Pick → *Query select the datum axis created in the previous step* → Angle → Plane → Pick → *Select ASM_TOP datum plane* → Done → Enter Value → 30 → ✓ **Refer Fig. 18.23.**
	35. Modify the angular dimension name.	Modify → Mod Dim → Dimension → Pick → *Click on ADTM1 and then select the angular dimension* → Done Sel → *Select the Dimension Text Tab* → (Name)CamAngle → OK → Done/Return **Refer Fig. 18.24.**

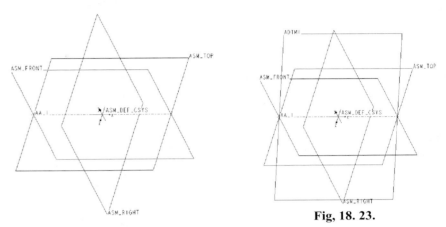

Fig. 18.22.

Fig, 18. 23.

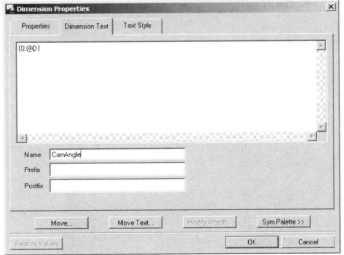

Fig. 18.24.

Goal	Step	Commands
Add relations to animate this datum plane	36. Add relations.	Relations → Assem Rel → Edit Rel → <u>Add relations as shown in Fig. 18.25</u> → FILE → SAVE → FILE → EXIT → Done
Verify the relationship	37. Verify the relationship.	Regenerate → Automatic → Repeat the regenerate command till the datum comes back to the original position. **Regenerate evaluates all the relations and, therefore, it increments the angle. If the cam subassembly is assembled to this datum, then it will also rotate with the datum plane.**
Assemble the cam subassembly	38. Start assembling the cam subassembly.	Component → Assemble → *Select cam.asm* → Open
	39. Move the assembly.	*Click on the move tab* → *Select the subassembly and move it* **Refer Fig. 18.26.**
	40. Align the axis of the shaft to the assembly datum axis.	*Click on the place tab →* (Constraint type) Align → (Component reference) *Select the axis of the shaft →* (Assembly reference) *Select the assembly datum axis*
	41. Align the subassembly TOP datum with the datum created in the step 33.	(Constraint type) Align → (Component reference) *Select the **subassembly** ASM_TOP datum →* (Assembly reference) *Select the **assembly** ASM_DTM1 →* ✓

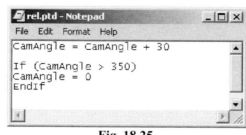

Fig. 18.25.

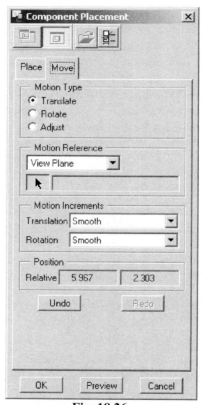

Fig. 18.26.

Goal	Step	Commands
Assemble the cam subassembly (Continued)	42. Align the subassembly RIGHT datum with the assembly RIGHT datum plane.	(Constraint type) Align → (Component reference) *Select the* **subassembly** *ASM_RIGHT datum* → (Assembly reference) *Select the* **assembly** *ASM_RIGHT datum plane* → 0.5 → ✅ **At this point, the assembly right datum plane should lie in the middle of the plate cam.**
	43. Place the cam subassembly.	OK → Done/Return **Refer Fig. 18.27.**
Verify the motion	44. Verify the motion.	Regenerate → Automatic
Suppress the cam subassembly	45. Suppress the cam subassembly.	Component → Suppress → Normal → Select → Pick → *Select the cam subassembly from the model tree* → Done Sel → Done
Assemble the left housing	46. Start assembling the housing part.	Assemble → *Select housing.prt* → Open
	47. Align the TOP datum with the assembly TOP datum plane.	(Constraint type) Align → (Component reference) *Select the TOP datum* → (Assembly reference) *Select the assembly TOP datum plane*
	48. Align the FRONT datum with the assembly FRONT datum plane.	(Constraint type) Align → (Component reference) *Select the FRONT datum* → (Assembly reference) *Select the assembly FRONT datum plane*

Fig. 18.27.

Goal	Step	Commands
Assemble the left housing (Continued)	49. Align the housing right (shelled) surface with the assembly RIGHT datum plane.	(Constraint type) Align → (Component reference) *Select the housing right surface* → (Assembly reference) *Select the assembly RIGHT datum plane*
	50. Place the housing part.	OK **Refer Fig. 18.28.**
Assemble the right housing	51. Start assembling the housing part.	Assemble → *Select housing.prt* → Open
	52. Align the TOP datum with the assembly TOP datum plane.	(Constraint type) Align → (Component reference) *Select the TOP datum* → (Assembly reference) *Select the assembly TOP datum plane*
	53. Mating the two-shelled surfaces of the housings.	(Constraint type) Mate → (Component reference) *Select the shelled surface of the right housing* → (Assembly reference) *Select the shelled surface of the left housing*
	54. Mate the FRONT datum with the assembly FRONT datum plane.	(Constraint type) MATE → (Component reference) *Select the FRONT datum* → (Assembly reference) *Select the assembly FRONT datum plane*
	55. Place the housing part.	OK **Refer Fig. 18.29.**

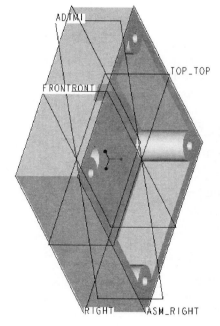

Fig. 18.28.

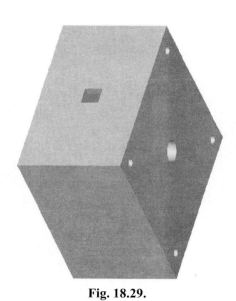

Fig. 18.29.

Goal	Step	Commands
Resume all features	56. Resume all features.	Resume → All → Done
Set up model display	57. Set up model display.	VIEW → MODEL SETUP → COMPONENT DISPLAY → Create → Vis0001 → ✔ → Hidden Line → Pick Mdl → Pick → *Select the two housings from the model tree* → Done Sel → Done → Done/Return **Refer Fig. 18.30.**
Suppress the housings	58. Suppress the housing.	Suppress → Normal → Select → Pick → *Select the housings from the model tree* → Done Sel → Done
Sketch a datum curve	59. Start "Datum – Curve" feature.	INSERT → DATUM → CURVE → Sketch → Done → Setup New → Plane → Pick → *Query select the* **assembly** *RIGHT datum plane (centered on the plate cam)* → Okay → TOP → *Query select the assembly TOP datum*
	60. Sketch the datum curve.	╲ → *Pick points 1 and 2* → *Middle Mouse* **Refer Fig. 18.31.**
	61. Modify the dimensions.	➤ → *Double click each dimension and enter the corresponding value* **Refer Fig. 18.31.**
	62. Exit sketcher.	✔

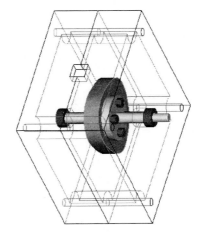

Fig. 18.30.

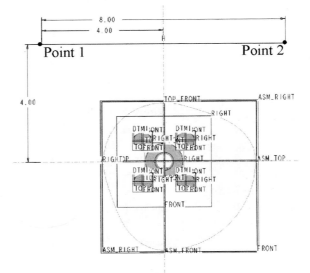

Fig. 18.31.

Goal	Step	Commands
Sketch a datum curve (Continued)	63. Accept the feature creation.	OK → VIEW → DEFAULT ORIENTATION **Refer Fig. 18.32.**
	64. Start "Datum – Curve" feature.	INSERT → DATUM → CURVE → Sketch → Done → Use Prev → Okay
	65. Sketch the datum curve.	☐ → *Loop → Select the four datum curves defining the cam profile*
	66. Exit sketcher.	✓ → VIEW → DEFAULT ORIENTATION
	67. Accept the feature creation.	OK
Create a datum point	68. Create a datum point.	DATUM → POINT → Crv X Crv → *Select the datum curve created in the step 66 and then, the straight line datum curve →* **Refer Fig. 18.32.**
Start creating the follower component	69. Start creating the follower component.	Component → Create → Follower → OK → **Refer Fig. 18.33.** Locate default datums → Three planes → OK

Fig. 18.32.

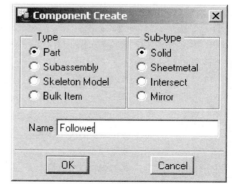

Fig. 18.33.

Goal	Step	Commands
Start creating the follower component (Continued)	70. Set up the datum planes.	Select → Pick → *Select assembly RIGHT datum plane* → Select → Pick → *Select assembly TOP datum plane* → Select → Pick → *Select assembly FRONT datum plane*
	71. Start "Protrusion – Extrude" feature.	Solid → Protrusion → Extrude → Solid → Done → Both Sides → Done
	72. Add a new reference.	*Select the datum point (APNT0) as a new reference*
	73. Sketch the section.	↘ → *Click to create the section shown in Fig. 18.34.*
	74. Add dimensions.	↦ → *Create the horizontal dimensions* **Refer Fig. 18.34.**
	75. Modify the dimensions.	↖ → *Double click each dimension and enter the corresponding value* **Refer Fig. 18.34.**
	76. Exit sketcher.	✔
	77. Define the depth and accept the feature creation.	Blind → Done → 1 → ✔ → OK

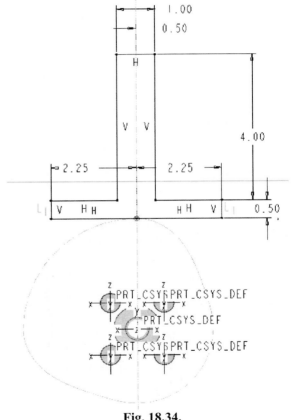

Fig. 18.34.

Goal	Step	Commands
Resume all features	78. Resume all features.	Resume → All → Done **Refer Fig.18.35.**
Verify the motion	79. Verify the motion.	Regenerate → Automatic **Repeat the process to identify any motion interference.**
Check the interferences	80. Check for the interferences.	ANALYSIS → MODEL ANALYSIS → (Type) Global Interference → Compute **ProE shows the volume interference between various components.** **Refer Fig. 18.36.**
Save the file and exit ProE	81. Save the file and exit ProE.	FILE → SAVE → CAMFOLLOWER.PRT → ✓ → FILE → EXIT → YES

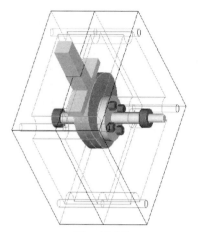

Fig. 18.35

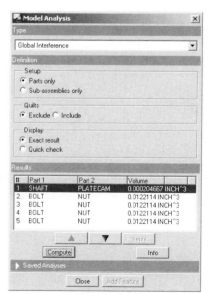

Fig. 18.36.

LESSON 19
ELECTRICAL FUSE ASSEMBLY

Learning Objectives:

- Learn to set up a *Layout* file.

- Practice creating parts in the assembly mode.

- Learn the use of *Data Tables*.

- *Import* from and *Export* to IGES files.

Design Information:

Sand-filled fuses are used to protect electric mains and feeders, circuit breakers, heating and lighting circuits, motors, transformers, semiconductors and more, against current surges. The components of a sand-filled fuse are: *Fuse element, End-caps, Fiberglass casing and sand.* Sand fills the space in the casing around the fuse element and helps in dissipating the energy during the short-circuit conditions. The cavity volume determines the amount of sand that can be filled. Therefore, this is one of the key parameters in the design. Several fuses are made with very similar geometry and a different number of weak spots. This lesson shows how to use layouts to control the model.

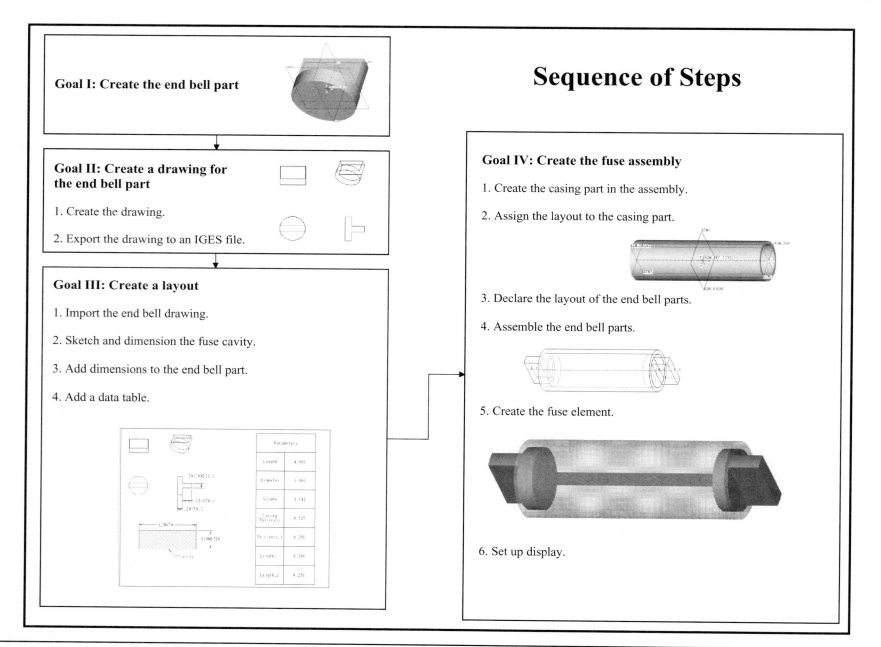

Goal I: Create the end bell part

Sequence of Steps

Goal II: Create a drawing for the end bell part

1. Create the drawing.

2. Export the drawing to an IGES file.

Goal III: Create a layout

1. Import the end bell drawing.

2. Sketch and dimension the fuse cavity.

3. Add dimensions to the end bell part.

4. Add a data table.

Goal IV: Create the fuse assembly

1. Create the casing part in the assembly.

2. Assign the layout to the casing part.

3. Declare the layout of the end bell parts.

4. Assemble the end bell parts.

5. Create the fuse element.

6. Set up display.

Goal	Step	Commands
Open a new file for the end bell part	1. Set up the working directory.	FILE → SET WORKING DIRECTORY → *Select the working directory* → OK
	2. Open new file for the end bell part.	FILE → NEW → *Part* → Solid → endbell → OK
Create the end bell part	3. Start "Protrusion – Extrude" feature.	Feature → Create → Solid → Protrusion → Extrude → Solid → Done
	4. Define the direction of extrusion.	One Side → Done
	5. Select the sketching plane.	Setup New → Plane → Pick → *Select the FRONT datum plane* → Okay
	6. Orient the sketching plane.	Top → Plane → Pick → *Select the TOP datum plane*
	7. Sketch a circle.	O → *Select the center of the circle at the intersection of TOP and RIGHT datum planes* → *Select a point to define the circle*
	8. Modify the dimensions.	⭦ → *Double click the diameter dimension* → 1 → **ENTER** **Refer Fig. 19.1.**
	9. Exit sketcher.	✔
	10. Define the extrusion depth.	Blind → Done → 0.25 → ✔
	11. Accept the feature creation.	OK → VIEW → DEFAULT ORIENTATION **Refer Fig. 19.2.**

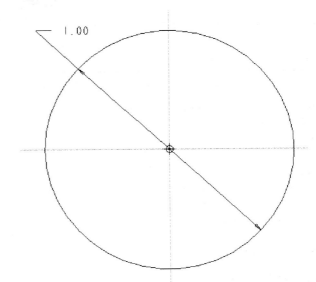

Fig. 19.1.

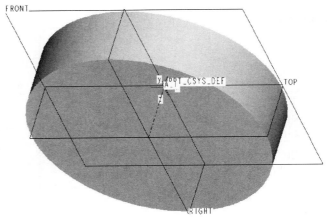

Fig. 19.2.

Goal	Step	Commands
Create the end bell part (Continued)	12. Start "Protrusion – Extrude" feature.	Create → Solid → Protrusion → Extrude → Solid → Done
	13. Define the direction of extrusion.	One Side → Done
	14. Select the sketching plane.	Use Prev → Flip → Okay
	15. Add new references.	*Select the circle* **Refer Fig. 19.3.**
	16. Create two lines.	↘ → *Pick points 1 and 2* → *Middle Mouse* → *Pick points 3 and 4* → *Middle Mouse* **Refer Fig. 19.4.**
	17. Dimension the lines from the TOP datum plane.	↔ → *Click on line 1 and the TOP datum* → *Middle Mouse* → *Click on line 2 and the TOP datum* → *Middle Mouse* **Refer Fig. 19.5.**
	18. Add a relation.	SKETCH → RELATION → ADD → <u>sd3 = sd4</u> → ✓ → ✓
	19. Use edge command to sketch a circle.	▢ → Loop → *Select the circle*
	20. Divide the circle at Points 1, 2, 3 and 4.	(icons) → *Select points 1, 2, 3 and 4*
	21. Delete arc segments 1 and 2.	➤ → *Select arcs 1 and 2* → ***DELETE***

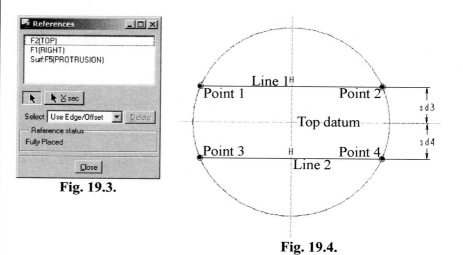

Fig. 19.3.

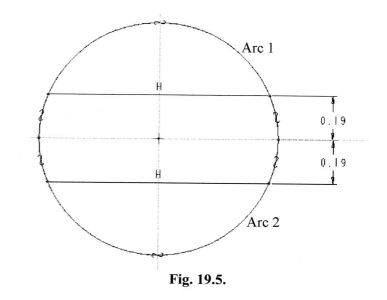

Fig. 19.5.

Goal	Step	Commands
Create the end bell part (Continued)	22. Modify the dimensions.	⬉ → *Double click the distance dimension* → 0.125 → ***ENTER***
	23. Exit sketcher.	✔
	24. Define the extrusion depth.	Blind → Done → 0.5 → ✔
	25. Accept the feature creation.	OK → VIEW → DEFAULT ORIENTATION **Refer Fig. 19.6.**
Save the end bell part	26. Save the part.	FILE → SAVE → ENDBELL.PRT → ✔
Open a drawing file for the end bell part	27. Open a drawing file for the end bell part	FILE → NEW → *Drawing* → ENDBELL → OK → (Default Model) ENDBELL.PRT → Use template → a_drawing → OK **Refer Fig. 19.7.**
Create the end bell drawing	28. Turn off datum planes.	⬛ *(Turn off datum planes and coordinate system)* → ▷
	29. Add a trimetric view for visualization.	Views → Add View → General → Full View → No Xsec → No Scale → Done → *Select the position for the general view* → (Type) Preferences → (Default Orientation) Trimetric → OK **Refer Figs. 19.8 and 19.9.**
	30. Move the views.	Move View → *Select the view* → *Select the destination position for the view* **Refer Fig. 19.9.**

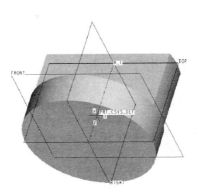

Fig. 19.6.

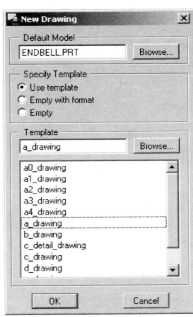

Fig. 19.7.

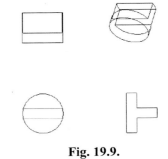

Fig. 19.8.

Fig. 19.9.

Goal	Step	Commands
Save and export the drawing	31. Save and export the drawing.	FILE → SAVE → ENDBELL.DRW → ✓ → FILE → SAVE A COPY → (Type) Iges → ENDBELL → OK
Open a new file for the fuse layout	32. Open a new file for the fuse layout.	FILE → NEW → *Layout* → fuse → OK **Refer Fig. 19.10.** *Select Landscape → Select A size* → OK **Refer Fig. 19.11.**
Import the end bell drawing	33. Append the end bell drawing to the model.	INSERT → DATA FROM FILE → *Select endbell.iges* → OK **Refer Fig. 19.12.**
	34. Move the drawing in the window.	EDIT → CUT → Pick Many → Pick Box → Inside Box → *Pick two points to define the diagonal of the selection box* → Done Sel → EDIT → PASTE → *Select the corner point of the drawing in the clipboard window → Select the destination position for the corner point* **Refer Fig. 19.12.**

Fig. 19.10.

Fig. 19.11.

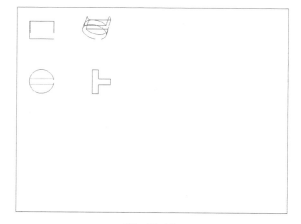

Fig.19.12.

Goal	Step	Commands
Sketch and dimension the fuse cavity	35. Create line 1.	SKETCH → CHAIN → SKETCH → LINE → SKETCH → SPECIFY ABS COORDS → (x) 1 → (y) 2 → ✓ → SPECIFY ABS COORDS → (x) 1 → (y) 3 → ✓
	36. Create line 2.	(x) 4 → (y) 3 → ✓
	37. Create line 3.	(x) 4 → (y) 2 → ✓
	38. Create line 4.	(x) 1 → (y) 2 → ✓ → ❌ **Refer Fig. 19.13.**
	39. **Optional Step:** If you make any mistakes, then delete excess lines.	↖ → Pick → *Select the lines to be deleted* → ***DELETE*** (to start new line creation)
	40. Hatch the cavity.	↖ → Pick → *Select the four lines* → EDIT → FILL → HATCHED → Cavity → ✓ → Spacing → Overall → Half → Half → Half → Done **Refer Fig. 19.14.**

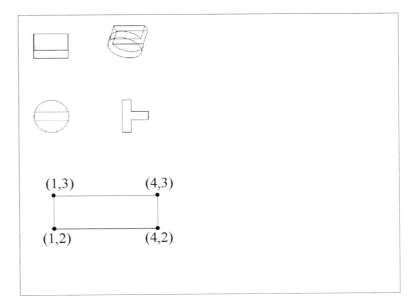

Fig. 19.13.

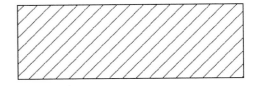

Fig. 19.14.

Goal	Step	Commands
Sketch and dimension the fuse cavity (Continued)	41. Dimension the cavity.	INSERT → DIMENSION → NEW REFERENCES → On Entity → Pick → *Select line 1* → *Middle Mouse* → Diameter → ✓ → 2 → ✓ → INSERT → DIMENSION → NEW REFERENCES → On Entity → Pick → *Select line 2* → *Middle Mouse* → Length → ✓ → 4 → ✓ → **Refer Fig. 19.15.**
	42. Name the cavity region.	INSERT → NOTE → Leader → Enter → Horizontal → Standard → Default → Make Note → On Entity → Arrow Head → Pick → *Select the bottom edge of the cavity (line 3)* → Done Sel → Done → Pick Pnt → *Select the starting point for the text note* → Cavity → ✓ → ✓ → Done/Return **Refer Fig. 19.15.**
	43. Add new parameters.	Relation → Add Param → Real Number → Volume → ✓ → 0.0 → ✓ → Real Number → Thickness_C (Casing thickness) → ✓ → 0.125 → ✓
	44. Add a new relation.	Add → Volume = pi * diameter^2 * length/4 → ✓ → ✓

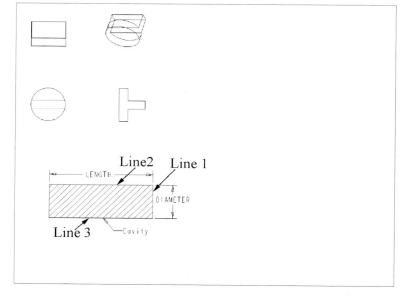

Fig. 19.15.

Goal	Step	Commands
Add dimensions to the end bell	45. Add dimensions to the end bell part.	INSERT → DIMENSION → NEW REFERENCES → On Entity → Pick → *Select thickness_1 line* → *Middle Mouse* → Thickness_1 → ✓ → 0.25 → ✓ → On Entity → Pick → *Select length_1 line* → *Middle Mouse* → Length_1 → ✓ → 0.5 → ✓ → On Entity → Pick → *Select thickness_2 line* → *Middle Mouse* → Length_2 → ✓ → 0.25 → ✓ → Return **Refer Fig. 19.16.**
	46. Arrange the dimensions.	↖ → Pick → *Select each dimension and move it to an appropriate location* **Refer Fig. 19.16.**
Add a data table	47. Create a table with eight rows and two columns.	Table → Create → Descending → Rightward → By Num Chars → Pick Pnts → *Select the upper right corner of the table* → *Mark off the width of the first column as 12 characters* → *Mark off the width of the second column as 12 characters* → *Middle Mouse* → *Mark off the height of the each row by clicking on 4 (Create 8 rows)* → *Middle Mouse* **Refer Fig. 19.17.**

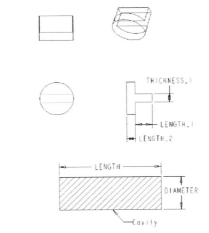

Fig. 19.16.

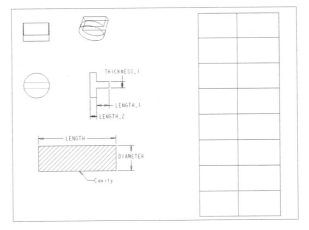

Fig. 19.17.

Goal	Step	Commands
Add a data table (Continued)	48. Merge the two cells in the first row.	Modify Table → Merge → Rows & Cols → *Select the two cells by clicking at the corners of each cell* **Refer Fig. 19.18.**
	49. Set alignment to center.	Mod Rows/Cols → Justify → Center → Middle → *Select the two columns*
	50. Enter the parameter names.	Enter Text → *Select each cell in the first column and enter corresponding parameter names (Split long names into two lines)* **Refer Fig. 19.18.**
	51. Define the parameter values.	Enter Text → *In the second column, enter the parameter names with the prefix "&" (example: &length for length) (&Thickness_C denotes the casing thickness)* → Done/Return → Regenerate **Refer Fig. 19.18.**
Save the layout	52. Save the layout.	FILE → SAVE → FUSE.LAY → ✓
Open a new file for the fuse assembly	53. Open a new file for the fuse assembly.	FILE → NEW → Assembly → Design → Fuse → OK
Declare the layout	54. Examine the existing relations.	Relations → Assem Rel → Show Rel → The relations window shows no relations → CLOSE → Done
	55. Declare the layout.	Set Up → Declare → Declare Lay → Fuse → ▼ → Done

Parameters	
Length	4.000
Diameter	2.000
Volume	0.000
Casing Thickness	0.125
Thickness_I	0.250
Length_I	0.500
Length_2	0.250

Fig. 19.18.

The volume is updated when the layout is regenerated.

Goal	Step	Commands
Declare the layout (Continued)	56. Examine the existing relations.	Relations → Assem Rel → Show Rel → The relations windows shows the relations from the fuse layout → CLOSE→ Done
Create the casing part in the assembly	57. Start creating the casing.	Component → Create → Part → Solid → casing → OK **Refer Fig. 19.19.** *Locate Default Datums → Three Planes →* OK **Refer Fig. 19.20.**
	58. Create the datum planes.	Select → Pick → *Select the ASM_RIGHT datum plane →* Select → Pick → *Select the ASM_TOP datum plane →* Select → Pick → *Select the ASM_FRONT datum plane → Select the casing part in the model tree to display its datum planes* **Refer Fig. 19.21.**

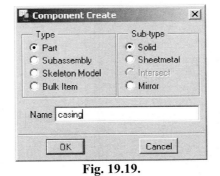

Fig. 19.19.

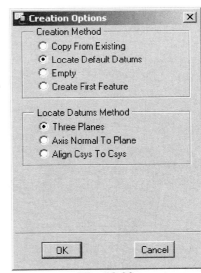

Fig. 19.20.

Fig. 19.21.

Goal	Step	Commands
Create the casing part in the assembly (Continued)	59. Specify the extrusion options.	Solid → Protrusion → Extrude → Thin → Done → Both Sides → Done
	60. Create a circle.	○ → *Select the center of the circle → Select a point to define the circle*
	61. Modify the dimensions.	↖ → *Double click the diameter dimension →* 1 → ***ENTER***
	62. Exit sketcher.	✔
	63. Select the direction of material creation.	Flip → Okay
	64. Define the thickness.	0.1 → ✔
	65. Define the depth.	Blind → Done → 4 → ✔
	66. Accept the feature creation.	OK → VIEW → DEFAULT ORIENTATION **Refer Fig. 19.23.**

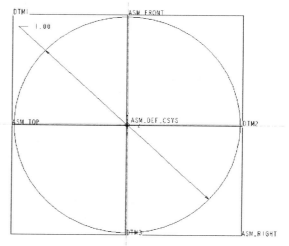

Fig. 19.22.

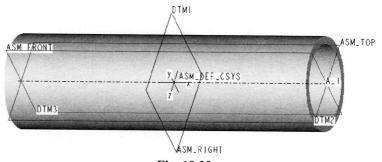

Fig. 19.23.

Goal	Step	Commands
Assign the layout for the casing part	67. Open the part in a new window.	*Right mouse on the casing part →* *Select Open* **ProE opens the part in a new window.**
	68. Declare the layout.	Declare → Declare Lay → Fuse
	69. Add relations.	Relation → Part Rel → Pick → *Select the protrusion →* **Refer Fig. 19.24.** Edit Rel → Input the relations shown in Fig. 19.25. → FILE → EXIT → YES → Done
	70. Regenerate the casing.	Regenerate (Note the changes.) WINDOW → FUSE.LAY → Advanced → Parameters → Modify → Diameter → 1 → ✓ → Done/Return → Regenerate → WINDOW → CASING.PRT → Regenerate (Note the link between the layout mode and part mode)

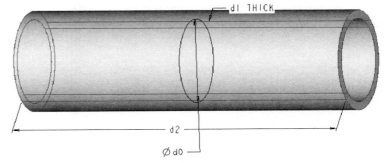

Fig. 19.24.

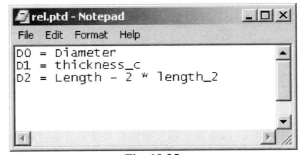

Fig. 19.25.

Goal	Step	Commands
Assign the layout for the end bell part	71. Activate the end bell part window.	WINDOW → ENDBELL.PRT
	72. Declare the layout.	Declare → Declare Lay → Fuse
	73. Add relations.	Relations → Part Rel → Pick → *Select the two protrusions* → **Refer Fig. 19.26.** Edit Rel → <u>Input the relations shown in Fig. 19.27. Note that d1, d2, d3 and d8 may be different for your model</u> → FILE → EXIT → YES → Sort Rel → Done
	74. Regenerate the end bell.	Regenerate
	75. Switch the active window back to assembly.	WINDOW → FUSE.ASM
Assemble the end bells	76. Start assembling the end bell.	Component → Assemble → (Look in) *In Session* → Select endbell.prt → Open **Refer Fig. 19.28.**

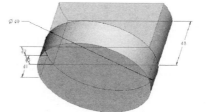

Fig. 19.26.

Fig. 19.27.

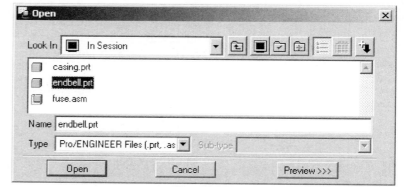

Fig. 19.28.

Goal	Step	Commands
Assemble the end bells (Continued)	77. Establish the constraints.	(Constraint Type) Align → Select → Query Sel → *Select the axis of the end bell* → Accept → Select → Query Sel → *Select the axis of the casing* → Accept → **Refer Fig. 19.29.** (Constraint Type) Align → Select → Query Sel → *Select the outer surface of the end bell (Refer Fig. 19.29)* → Accept → Select → Query Sel → *Select the outer surface of the casing (Refer Fig. 19.29)* → Accept → 0 → ✓ → OK → Done/Return **Refer Fig. 19.30.**
	78. Place the second end bell.	Adv Utils → Repeat → Select → Pick → *Select the end bell part from the model tree* → *Select the "Align Surface Surface" constraint* **Refer Fig. 19.31.** ADD → Select → Pick → *Select the other end surface of the casing (highlighted in Fig. 19.32)* → Confirm → ⊞ → Done/Return → Done/Return **Refer Fig. 19.33.**

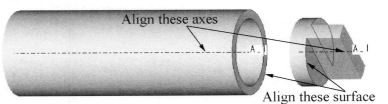

Align these axes

Fig. 19.29.

Align these surface

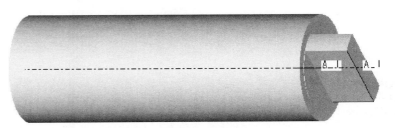

Fig. 19.30.

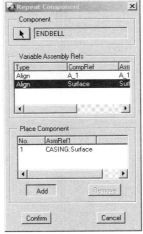

Fig. 19.31.

Fig. 19.32.

Fig. 19.33.

Goal	Step	Commands
Create the fuse element	79. Start creating the fuse element.	Component → Create → *Part* → *Solid* → element → OK → *Locate Default Datums* → *Three Planes* → OK
	80. Define the datum planes.	Select → Pick → *Select the ASM_TOP datum plane* → Select → Pick → *Select the ASM_FRONT datum plane* → Select → Pick → *Select the ASM_RIGHT datum plane* → *Select the fuse element part in the model tree to display its datum planes* **Refer Fig. 19.34.**
	81. Start "Protrusion – Extrude" feature.	Solid → Protrusion → Extrude → Solid → Done
	82. Define the direction of material creation.	Both Sides → Done
	83. Add new references.	*Select the right end of the left end bell and the left end of the right end bell* **Refer Fig. 19.35.**
	84. Sketch the section.	▢ → *Pick points 1 and 2* **Refer Fig. 19.35.**
	85. Modify the dimensions.	↖ → *Double click the width dimension* → 0.25 → ***ENTER*** → *Double click the half-width dimension* → 0.125 → ***ENTER***
	86. Exit sketcher.	✔
	87. Define the depth.	Blind → Done → 0.02 → ✓

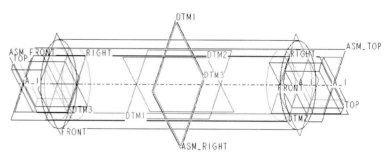

Fig. 19.34.

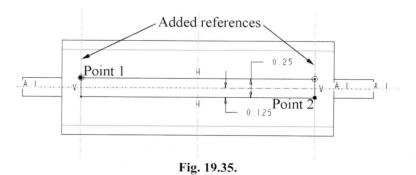

Fig. 19.35.

Goal	Step	Commands
Create the fuse element (Continued)	88. Accept the feature creation.	OK → VIEW → DEFAULT ORIENTATION **Refer Fig. 19.36.**
Set up color	89. Define the model color.	VIEW → MODEL SETUP → COLOR AND APPEARANCES → ADD → *Click on the white area →* **Refer Fig. 19.37.** *Click on Color wheel → Select a suitable color in the color editor window →* ADD *in the Appearances window → Add three more colors →* OK *→ Select a color →* Component → Pick *→ Select the components →* Done Sel → Set → Set the colors for all the components except the casing → *Select a suitable color for the element →* ADD *→ Click on the Advanced tab → Slide the transparency tab →* OK *→* Component → Pick *→ Select the casing →* Done Sel → Set → Close *→* 🔲 **Refer Figs. 19.38 and 19.39.**
Save the file and exit ProE	90. Save the file and exit ProE.	FILE → SAVE → FUSE.ASM → ✔ → FILE → EXIT → YES

Fig. 19.36.

Click HERE

Fig. 19.37.

Fig. 19.38.

Fig. 19.39.

LESSON 20
TRICKS OF THE TRADE

Chamfers

The chamfer command can be accessed by:

Feature → Create → Chamfer.

The user can chamfer an edge or a corner using the chamfer command. The options under the chamfer - edge command are:

- 45 x d
- d x d
- d1 x d2
- Ang x d

The corner can be removed by specifying either points or distances along each of the lines creating the corner.

Refer Fig. 20.1.

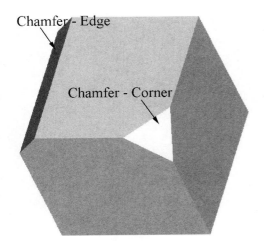

Fig. 20.1.

Configuration file

Configuration options specify how ProE handles the modeling tasks, and also the display. The file allows the user to specify the display options such as color, and modeling options such as the number of digits and tolerances. The config.sys file can be accessed by:

UTILITIES → OPTIONS

It opens the preferences window (Refer Fig. 20.2). The user can search for the desired option using FIND. This configuration file can be used to turn-off the annoying BELL sound. The configuration file can be saved. The user may create his/her own configuration file and open it from this window. When ProE is started, it looks at config.sup (system configuration file) and then, configuration files in the ProE loading point, user login and working directories. The last value of the configuration option will be used.

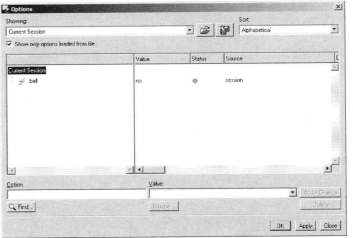

Fig. 20.2.

Cosmetic features

Sketched cosmetic features include cosmetic features such as company name, part number and other manufacturing information. They are also useful in creating regions so that loads can be applied later in ProMechanica. The cosmetic feature can be created by the following procedure:

Feature → Create → Cosmetic → Sketch →

Regular Sec/Proj Sec →
Projected section projects the section drawn onto a surface.

Xhatch/No Xhatch
This option allows hatching of the cosmetic feature.

Select the sketching plane → Okay →
Top → *Select the top surface to orient the sketching plane* →

SKETCH → TEXT → *Select the start point of the text* → *Select another point to determine the text height and orientation* →

Refer Fig. 20.3.

(Text line) <u>Enter Text</u> → *Select the font, aspect ratio and slant angle* → ✔ →

↖ → *Modify the dimensions* → ✔ → Done

Refer Fig. 20.4.

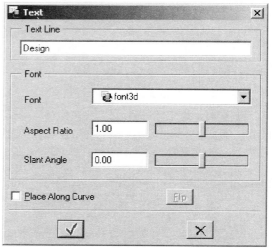

Fig. 20.3.

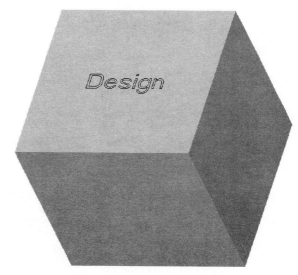

Fig. 20.4.

Mapkey

Mapkeys is a command to create macros for frequently used command sequences. For instance, the mapkey command allows the user to change the view to the default orientation, and then shade the model with a single keystroke. Mapkey can be defined by the following set of commands:

UTILITIES → MAPKEY

ProE opens the mapkey window.
NEW → CLOSE

In the record mapkey window:

(Key Sequence) Type the keyboard key name. For function names use prefix $. (For instance, type $F1 for F1) →

(Name) Type a name for the macro → (Description) Type description → RECORD → Select the sequence of commands (For instance, VIEW → DEFAULT ORIENTATION → VIEW → SHADE) → STOP → OK → CLOSE

Mapkey can be added to the screen using the following set of commands:

UTILITIES → CUSTOMIZE SCREEN → Select the COMMAND tab →

(Categories) Select Mapkey → Right Mouse → Choose button image → Double click the image → Drag the mapkey button to the desired menu and release it → OK

Material specification

Material properties can be defined using the following command:
Set Up → Material → Define
(Enter the material name) Name of your material → ✓ →
Enter the values of the material in the material editor (Refer Fig. 20.5).

Fig. 20.5.

FILE → SAVE → FILE → EXIT
Once defined, material can be assigned to a part by the following commad:
Set Up → Material → Assign → From Part →
Select the defined material → Accept

Weight and model information can be obtained using:
Analysis → Model Analysis → (Type) Model Mass Properties →
COMPUTE → CLOSE

Rounds

Simple rounds can be created by the following commands:

Select the edge(s) to be rounded → *Right Mouse* → *Round* →
*Drag the handles to modify the radius of the round or double clicking the
radius dimension and entering corresponding value*

A round with variable radii (Refer Fig. 20.6) can be created between
two surfaces by using the following commands:

INSERT → ROUND → Simple → Done →
Variable → Edge-Chain → Done
Tangent Chain → Pick → *Select the edges to be rounded* → Done Sel
→ Done →
(Optional Step) *Select intermediate points where the radius needs to
be specified* → Done →
Specify the radius at different points → OK

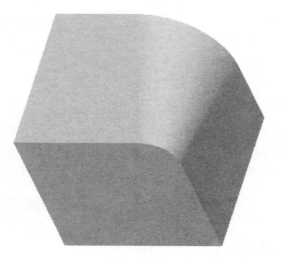

Fig. 20.6.

Trail.txt files

ProE records all the commands, menu selections used, and dialog
choices in a file called "trail.txt." This file can be used to either re-create
a session or create training files. Note that the file should be renamed
before opening it in ProE. This file is editable in a text editor. The user
can edit the file to change the options. A trail file can be played in ProE
by the following command:

UTILITIES → PLAY TRAIL/TRAINING FILES

INDEX

Sketcher Mode